AF657431

Marine Fisheries Ecosystem: Its Quantitative Evaluation and Management

Marine Fisheries Ecosystem: Its Quantitative Evaluation and Management

A.C.Long

CYBER TECH PUBLICATIONS
4264/3, Ansari Road, Daryaganj, New Delhi-110002 (India)
Ph.: 011-23244078, 011-43559448 Fax: 011-23280028
E-Mail: cyberpublicationsdelhi@yahoo.com
Website: www.cybertechpublications.com

MARINE FISHERIES ECOSYSTEM: ITS QUANTITATIVE EVALUATION AND MANAGEMENT

A.C.Long

© Reserved

First Edition 2010

All rights reserved. No part of this publication may be reproduced, stored in a retrieval system or transmitted, in any form or by any means, electronic, mechanical, photo copying, recording or otherwise, without the prior permission of the publisher.

Published by Asian Offset Printers

G.S. Rawat for **Cyber Tech Publications**
4264/3, Ansari Road, Daryaganj, New Delhi-110002 (India)
Ph.: 011-23244078, 011-43559448 Fax: 011-23280028
E-Mail: cyberpublicationsdelhi@yahoo.com
Website: www.cybertechpublications.com

Preface

Fisheries have been important parts of human life and food production throughout history. Fisheries have become a part of human cultures and mythologies, providing a community identity and a subject for artists throughout the ages. Partially, this is because fisheries are irretrievably wrapped up in humanity's perpetual fascination with the sea, and partially, because they have been a major source of food and income for many communities throughout the ages. Today, fisheries are estimated to provide 20 per cent of the world population's protein, and that figure is considerably elevated in some developing nations and in regions that depend heavily on the sea. Fisheries are a huge global business and provide income for millions of people.

The large natural river supply of phosphorus and nitrogen, essential nutrients for marine plants and algae. The tiny floating marine plants known as phytoplankton which form the base of the marine food chain are either eaten or die and gradually fall to deeper waters where bacteria take care of decomposing them, almost completely. Replenishment of the bottom waters of the sea with new seawater from the Mediterranean takes hundreds of years. Despite this situation, for millennia, its surface waters supported a rich and diverse marine life.

This book provides a global overview of the historical trends and of the present situation of the marine fisheries sector. It offers a global picture of the state of the marine fish resources, and an elaborate description of the fishery system-- the technologies, production, trade and its contribution to food security. It also discusses the issues of marine fisheries governance and the trends towards ecosystem-based management. The book will be of interest to students, fisheries scientists, people engaged in small-scale/large-scale fisheries and policy planners involved with fisheries management and poverty alleviation

—Author

CONTENTS

1

Introduction

The “systems approach” recognizes the diverse factors affecting aquaculture and is a multifactorial and multidisciplinary approach that attempts to analyse how different factors affect aquaculture and develop solutions to problems based on an understanding of how aquaculture systems operate. The systems approach is fundamentally a multi-disciplinary approach that can be used to solve problems and identify opportunities for development.

This analytical approach has been shown to contribute to identification of key research issues, development of better management solutions, improvement of business efficiency, design and testing of new aquaculture systems and more effective extension and education, among others. This paper discusses the systems approach to aquaculture, provides examples of the relevance and use of the approach in aquaculture development, and recommends areas for further study and follow up actions.

The varied topics covered in these Proceedings of the Conference on Aquaculture in the Third Millennium, from highly technical sessions on genetics and feeds, to nontechnical policy sessions on the role of different stakeholders and institutions, reflect the wide range of factors and issues relevant to modern aquaculture development.

The “systems approach” recognizes this diversity of influences on aquaculture development, and is a multifactorial and multidisciplinary approach. It uses an understanding of how aquaculture systems operate to analyse how different factors affect aquaculture and develop solutions to problems that are identified. This analytical approach has been shown to contribute to identification of key researchable issues (Edwards, 1998; Smith, 1999), development of better management solutions, improvement of business efficiency, and design and testing of new aquaculture systems, as well as to more effective extension and education.

The analysis and understanding of aquaculture systems can occur at different "levels":

* the organism and its surroundings;
* the production unit;
* the economic enterprise;
* the farm, watershed or coastal areas;
* the national sectoral level, or even
* the international level.

The boundaries chosen for the system of interest may be physical entities, political borders or organizational structures. In terms of this paper, the focus is on the following systems: the farm; the farm and its local environment; and the national level, recognizing the fact that there are critical interactions between these different levels.

This paper is intended to be complementary to the earlier papers in this technology session which focused on the individual components of the system-such as feed and nutrition, health, seed and genetics-and emphasises understanding of how these individual components interact within the wider context of aquaculture systems. It first looks at our current understanding of aquaculture systems and management strategies based on a systems approach, with an emphasis (as in this session) on farm-level management practices.

As social, economic, policy and institutional issues are all important-at some level-in a systems approach, the paper also provides a link between the Conference's farming technology and policy sessions and closes with some suggestions for discussion on future directions. This paper was developed from discussions and contributions made by panel members and participants during the Conference on Aquaculture in the Third Millennium. The recommendations presented here represent a consensus of recommendations adopted in the final plenary session of the Conference.

AQUACULTURE SYSTEMS AND SUSTAINABILITY

Sustainable development and sustainability are complex issues that are difficult to define and apply to aquaculture. The "systems approach", however, can assist understanding of these issues, as they relate to aquaculture development.

The term sustainability has been defined in various ways but perhaps the most widely used is based on the definition of "sustainable development" in the Brundtland report: "sustainable development is development that meets the needs of the present without compromising the ability of future generations to meet their own needs". An even more succinct definition is that of the International Union for the Conservation of Nature (IUCN): "sustainable development improves people's quality of life within the context of the Earth's carrying capacity". These definitions contain two key concepts: meeting the present and future needs of the world's people, and accepting the limitations of the environment to provide resources and to receive wastes for the present and for the future. The Food and Agriculture Organization of the United Nations (FAO), in particular, recognized that increased capacity at the national level is required to achieve sustainable development by including the need for "institutional change" in definitions of sustainable agricultural development (FAO, 1995). The recognition that institutions are important highlights the need for education and training, effective institutional arrangements and a legal and policy framework to underpin sustainable development of agriculture, and indeed aquaculture.

Sustainability is commonly split into three separate components: social sustainability (SS); economic sustainability (EcS); environmental sustainability (ES). Whilst social sustainability criteria are difficult to define, the definition of economic and environmental sustainability are providing a basis from which management options can evolve in aquaculture projects. To take sustainability to a more practical level requires consideration of environmental, social and economic issues in aquaculture development. Thus, the approach to sustainability implies a systems approach.

There are general guidelines available on the different issues to consider. The Code of Conduct on Responsible Fisheries (CCRF), adopted by the FAO Conference in 1995 (FAO, 1995) in particular, identifies a number of key issues. The Code sets out principles and international standards of behaviour for responsible practices, to ensure effective conservation, management and development of living aquatic resources while, at the same time, recognizing the nutritional, economic, social, environmen-tal and cultural importance of fisheries and aquaculture, and the interests of all those involved in these sectors. The Code's Article on Aquaculture Development (Article 9) contains provisions relating to aquaculture, including culture-based fisheries. Fundamentally, the Code recognizes the importance of activities that support the development of aquaculture at different levels:

* the producer level;
* the local area, i.e., the farm and its integration into local area management and rural development schemes;
* the national institutional and policy environment; and
* international and transboundary issues.

The Code identifies many key principles in development of management strategies based on an understanding of aquaculture systems-from the farm to national and international levels. It also provides a basis for a systems approach.

THE SYSTEMS APPROACH

Farm Level

There is a lot of information on aquaculture farming systems, and various definitions are available, such as the level of intensity of management and output, and degree of integration with other on-farm activities. A considerable literature exists on integrated (agriculture-aquaculture and vice versa) farming systems (Edwards, 1998 and "Farming species and systems" in these proceedings).

However, there are a wide range of culture species, culture facilities and management practices in use, and thus a very wide range of farming systems.

Key factors to be understood in the functioning of a farming system are the technologies of production and social, economic and environmental aspects. At the technology level, feeds, feed additives and fertilizers, water quality, seed quality and availability, chemoterapeutants and other chemicals, disposal of wastes that may adversely affect human health and/or the environment, and food safety of aquaculture products all require consideration. It has also been emphasised that a better understanding of the microbial populations in aquaculture systems, their interaction with the health of the farmed animals, and their role in maintaining a healthy aquatic environment are required.

The systems approach at the farm level can be used to understand and improve the efficiency of use of key natural resources-particularly water, nutrients, land, seed and financial resources. When focussing on the mix of all sustainability criteria through a systems approach, we have to deal with such factors as: appropriate densities, production and husbandry systems geared to the animal's health, maintaining ecological balance within the pond or other growout habitat, and provision of optimal social and economic benefits. It is inevitable that impacts on natural resources will become an increasingly important issue in the new millennium, and the systems approach can be used to analyse and develop the solutions required for more sustainable use of natural resources.

A wide range of management systems is already employed in aquaculture operations with varying degrees of success. Given that aquaculture systems range from small, relatively self-contained farms for subsistence, to large-scale commercial units for trade purposes, variable success is hardly surprising. Thus, a "one-size fits all" approach is unlikely to be successful. A systematic approach to production management, however, allows the farmer to manipulate and control production inputs that will result in more efficient, cost-effective production and minimize excessive outputs with negative environmental impacts. There is a tremendous body of information on site selection, farm construction and design features, aquatic animal health management, broodstock and seed production and care, production techniques, the use of appropriate feeds, feed additives and fertilizers, water and sediment management, including effluent control, and other topics (e.g., Chanratchakool et al. 1998; FAO/NACA, 1995; Pennell and Barton, 1996).

The challenge is to optimize dissemination and use of such information and experience.

The systems approach can also be used for aquatic animal health management. Here the emphasis needs to move more towards management procedures, policies and products that can prevent or effectively eradicate significant pathogens, prevent re-infection through contaminants, and manage diseases in an environmentally sustainable manner. Subasinghe et al. (1998) provide a discussion of the role of the systems approach in aquatic animal health management.

The Local Level

The systems approach at the local level recognizes that individual farms cannot be seen in isolation, and that there are many interactions between an aquaculture farm and the external environment-including environmental resources and local communities. Furthermore, there can be significant cumulative effects where there are large numbers of farms crowded in small areas.

Environmental interactions with aquaculture arise from a wide range of inter related factors including availability, amount and quality of resources; type of species cultured; size of farm; culture systems management; and environmental characteristics of the farm location. Environmental interactions are not limited to impacts of aquaculture on the environment, but include environmental impacts on aquaculture and impacts of aquaculture on aquaculture. Perhaps less well known or documented are the many ways that aquaculture can contribute to environmental improvement, for example mollusc farms' desedimentation or improved nutrient turnover (Hatcher et al., 1994), or water storage on small-scale freshwater farms.

At the local level, social and institutional interactions are also important and need to be better understood, for example:

* participation of, and benefits to, rural communities;
* institutional support through extension services;
* access to information etc.

A systems approach attempts to understand these linkages and develop management strategies based on such understanding.

The future of integration of aquaculture into local ecological and social systems requires more focus on local area development planning. Fortunately, increasing attention is being given to such issues, particularly in coastal areas. Integrated coastal management (ICM) is a process that addresses the use, sustainable development and protection of coastal areas, and according to GESAMP (1996) "comprehensive area-specific marine management and planning is essential for maintaining the long-term ecological integrity and productivity and economic benefit of coastal regions". ICM is made operational through such activities as:

* land use zoning and buffer zones;
* regulations, including permitting to undertake different activities;
* nonregulatory mechanisms;
* construction of infrastructure;
* conflict resolution procedures;
* voluntary monitoring; and
* impact assessment techniques.

More participatory approaches to planning of aquaculture development will also be given attention with the move towards integrated development planning (see S. Sen, this volume). Practical experience in implementation ICM for aquaculture is limited, which is in large measure because of the absence of adequate policies and legislation and institutional problems, such as a lack of unitary authorities with sufficiently broad powers and responsibilities, as well as limited training and education of people concerned.

In inland rural areas, increasing attention is being given to integration of aquaculture into rural development and special area management plans. Increasing emphasis is also being given to promotion of aquaculture for poverty alleviation (see Haylor and Bland, this volume). Such an approach requires emphasis on immediate social needs and people's livelihoods (and how aquaculture might meet these needs and contribute to improved livelihoods) rather than a technology/aquaculture driven approach. The emphasis on aquaculture for development, rather than development of aquaculture may lead to some fundamental changes in the approach to promotion of aquaculture in the coming years.

National, Regional and International Levels

At the national level, government policy, and institutional and human capacity are most important in providing a strong foundation for aquaculture to develop in a su (Insull and Shehadeh, 1996).

These issues are covered extensively in the Conference's policy session and are not discussed in detail here, except by recognition that most community and farm activities are influenced by national-level policy, legislation and institutional support. For example, the level of aquatic animal disease affecting small-scale producers or enhanced fisheries is related to national policies for quarantine and movement of live aquatic animals, which affect the risk of exposure of small-scale producers to serious aquatic pathogens (DFID/FAO/NACA/GoB, 2001). Inter-and intra-country trading patterns and movement of aquatic animals also affect these risks.

International conventions (e.g. the Convention on Biodiversity), trade and consumer preferences, all clearly impact aquaculture development at a local level.

APPLICATION OF THE SYSTEMS APPROACH

Some examples of systems approaches presented during the Conference are given below.

Small-scale Farmer Research and Extension

The systems approach has been widely used in the promotion of sustainable development through small-scale integrated aquaculture (Edwards, 1998). He emphasises that most scientists focus on technical aspects of aquaculture, resulting in the impression that the major constraint facing aquaculture development is a shortage of technical knowledge, overshadowing the developmental and educational constraints.

The most important constraint to aquaculture development is dissemination of existing knowledge, whether derived from research or indigenous technical knowledge of farmers also affect these risks. The limited capacity of developing-country institutions in education, research and development compounds this fundamental failing. Research should follow farming systems research and extension methods in which inter-disciplinary teams work with farmers to evaluate and develop both production systems and extension methods that are appropriate to the local conditions of farmers and their resource base.

A systems approach to management has more potential for success on the typically diverse small-scale farms in developing countries. This approach includes analysis of the resource-base of the farm, and the farmers' perceptions of their needs.

The role of institutions in promoting the development of aquaculture can also be analysed using such an approach. The role of national institutions (government and non governmental agencies) as primary facilitators of aquaculture and the necessity of capacity building in order to facilitate the systems development approach should be recognized. International and regional agencies can support such capacity building.

ENVIRONMENTAL MANAGEMENT SYSTEMS

Environmental management systems are also starting to evolve for some forms of aquaculture, as well as more formal environmental management systems (EMS). EMS is a complex approach with little direct application to small-and medium-sized farms, however, EMS principles provide useful guidance for improving environmental management of aquaculture production systems (H. Dixon, pers. comm.).

EMS assemble management policies, programmes and practices designed to identify links between industry, urban and developmental activities, and consequent pressures on the environment. An effective EMS for aquaculture should establish indicators of changes in the environment, including land, water and aquatic resources. Policies and practices responding to the changes are implemented with continuous feedback to reduce/ mitigate any environmental impacts. These indicators should make clear links between environmental impacts and aquaculture activities.

The indicators should also reflect positive and negative impacts of environmental influences on aquaculture, as well as visa versa. EMS provides a possible systematic approach to motivate aquaculture to better organize priorities and projects to identify problems and potential impacts before they occur, as well as meet environmental and business goals. This process also assists compliance with national environmental laws and regulations.

A successful EMS provides the means by which aquaculture can identify causes of environmental problems and prevent them, thus saving money to repair or mitigate after the damage has been done.

According to H. Dixon, key benefits of an EMS include:

* improved environmental performance,

* reduced liability,

* competitive advantages,
* improved compliance,
* reduced operational costs,
* enhanced consumer trust, and
* increased access to capital.

H. Dixon considers that formulating an institutional framework for environmental management systems of aquaculture should:

* be simple to implement, clear and comprehensible to all involved, including the public;
* consider the needs of all stakeholders;
* consider individual abilities and resources;
* be financially flexible and not inhibit the activities it is designed to address; and
* be based on sound scientific information, quantifiable and effective.

CODES OF PRACTICE

The technical methods, management systems and practices needed for minimizing impacts are being increasingly incorporated into more formal "Codes of Practice", notably in more commercially oriented and intensive shrimp and salmon farming. Several aquaculture organizations, for example, the Marine Shrimp Culture Association of Thailand, the Global Aquaculture Alliance, the Australian Prawn Producers Association, the Irish Salmon Growers Associations and others have taken the FAO Code of Conduct a step further and formulated Codes of Practice (COP). These COP contain principles for preventing or mitigating negative environmental and social impacts through use of "best management practices" (BMP).

They are currently for voluntary adoption and consist of documented guidelines available to farmers. Much needs to be done towards their implementation, such as the development of operational manuals and support programmes. Furthermore, the extent to which COP will be fully adopted by farmers under self-regulation and the environmental consequences of their adoption remain to be determined. Implementation is a very important issue for the new millennium, particularly for small-scale farmers.

Improving Profitability of Aquaculture Operations

The systems approach is directly applicable to the business of aquaculture, through the use of a structured, systematic approach to operations and business management.

E. Hempel considers that in high technology salmonid aquaculture, the "software" of aquaculture (people) is more important than the hardware or "technology". There are many facts and variables, so in order to understand and manage the system, things have to be simplified-"systematized"-. The approach emphasised by E. Hempel involves:

* understanding and identifying the system of concern within the farm enterprise/business;
* simplifying the system by identifying only key system variables;
* for each person working at different levels of the organization, identifying the key system variables to monitor;
* reducing the number of variables to the minimum-e.g. dissolved oxygen, food conversion ratio (FCR), cash flow, and set limits/standards; and
* monitoring and responding to changes that exceed the set limits.

E. Hempel also emphasises that effective management based on a systems approach requires assigning of responsibilities within an aquaculture business to monitor and respond to changes that occur.

He also emphasises that the system may change with time, and that management has to be flexible to respond to such changes. He further considers that the systems approach to business and management is important to improving the performance and profitability of farming, and is relevant at all levels of aquaculture, from small-scale farms to the largest business.

Institutional use of a Systems Approach

A systems approach can also be used to define institutional responsibilities in aquaculture. For example, in western Australia, according to C. P. Rogers' presentation during the "Systems session" of the Conference on Aquaculture in the Third Millennium, NACA/FAO, 2000) a systematic analysis of institutional responsibilities was useful to establish the key points for decision making in allocation of land and water resources for aquaculture development. This helped streamline permit processing. He also emphasised that a systematic evaluation of management processes can be valuable in identifying institutional responsibilities at both government and private-sector levels, as well as for promotion of small-scale aquaculture.

Implications of Systems Approach for Future Aquaculture Development

Aquaculture can be socially, environmentally and economically sustainable and, contribute to the production of food and rural development, provided appropriate farming systems and management practices are adopted.

The multidisciplinary and multisectoral systems approach recognizes that technical, economic, social and environmental issues, as well as institutional factors, have to be considered in the process of development and management of aquaculture. The systems approach attempts to understand the way the system operates and the interactions between different components, and serve as a basis for better management. This multidisciplinary approach requires different skills and, as such, also needs cooperation among different disciplines and information exchange among different stakeholders. The application of a systems approach to aquaculture has several implications for a more analytical and structured approach to aquaculture development.

There is a need for better information on aquaculture systems and promotion of more effective information exchange between stakeholders.

Information requirements include social and environmental interactions at the farm level, development of better practices targeting important environmental and social impacts, and seeking incentives for farmers to adopt better farming practices.

Development of suitable indicators is also another important factor. For assessing natural resource use by aquaculture, recent research has emphasised the "ecological footprint" (Kautsky et al., 1997). The aims of the authors to develop a "sustainability" index are to be commended, however, there are a number of problems with the models promoted so far (Roth et. al., 2000), including:

* their static and dimensional (area) nature;
* the difficulties of incorporating economic and social values into the approach;
* the emphasis on biophysical factors;
* the lack of the footprint and ecological capacity to accept multisectoral or multiple uses;
* the lack of inclusion of ecosystem services (such as absorption capacity for discharge of nutrients into coastal ecosystems); and
* difficulties in making comparisons between different locations and systems.

There was a consensus in the Conference discussions on the need to develop and evaluate meaningful indicators of resource use efficiency to better define environmental interactions with aquaculture and improve management decisions. Research is required to develop and evaluate meaningful indicators of resource use efficiency useful to management decisions.

Institutional strengthening in the private and public sectors is another critical issue that must be addressed. The systems approach recognizes that national institutions (government and non governmental agencies) are the primary facilitators of aquaculture development in their countries.

However, national institutions involved in the promotion of aquaculture need to increase their capacity and reinforce emphasis on multidisciplinary approaches. This will require a shift away from traditional "top down" and "technology focused" capacity building and extension approaches, towards an approach that is interactive and responsive to farmer needs.

As emphasised above, much aquaculture research to date has focused on technical questions, leaving the impression that the major constraint facing aquaculture is a shortage of technical knowledge.

This detracts from the attention needed for development and basic aquaculture education (Edwards, 1998). Another important constraint to aquaculture is dissemination of existing knowledge, whether derived from research or the indigenous technical knowledge of farmers. The limited capacity of many developing countries' national institutions in education, research and development concerning the promotion of aquaculture compound this problem, and require serious attention in the new millennium.

Examples of the systems approach mentioned during the panel discussions included:

* its application to small-scale farmer research and extension;
* the development and application of best management practices to improve environmental management on commercial aquaculture farms;
* improving the profitability of aquaculture operations through the use of a structured, systematic approach to operations and business management; and
* the institutional use of a systems approach to assist in effective decision making at a regulatory/planning level and within the aquaculture industry as a whole.

RECOMMENDATIONS OF THE BANGKOK CONFERENCE

The Following Recommendations were Adopted by the Millennium Conference

That a multidisciplinary and multisectorial and systematic approach be taken in the development of aquaculture and aquaculture research. Applying a systems approach allows the proper understanding and analysis of problems and opportunities, and the development of solutions based on the understanding of how systems operate.

The systems approach involves basically seven steps:

* state the problem;
* identify the system;
* classify and describe the system, boundaries and key factors;
* analyse the problem;
* propose solutions;
* test the solution; and
* implement and disseminate knowledge to solve the problem.

Codes of Practice, and the Best Management Practices incorporated into such codes offer an opportunity to improve management.

These need to be developed further and applied in real situations as a practical basis for the operating system. Special attention needs to be given to:

* implementation of Codes of Practice;
* the need for continual fine tuning to take account of new information and technology;
* a role for farmers in implementation (there should be greater participation by farmers and other end-users to provide feedback and fine-tune systems management practices); and
* the effectiveness of self-regulation.

There was a consensus on the need to develop and evaluate meaningful indicators of resource use efficiency to assist in understanding environmental interactions with aquaculture and improve management decisions. The use of the "ecological footprint" model as a tool was discussed, and it was suggested to develop a more appropriate index.

There should be more emphasis on multidisciplinary research using a systems type approach, to better understand problems and opportunities and develop more appropriate solutions to overcome aquaculture development constraints and unfulfilled potential.

There is a need to communicate practical examples where the systems approach has been used and to incorporate the problem identification and solving systems approach into educational and training programmes.

2

Aquaculture Products

The growth of aquaculture has led to significant changes in how its products are perceived and marketed. In becoming an important contributor to the markets for seafood, aquaculture is increasingly subject to safety mechanisms and controls, such as the statutory hazard analysis critical control point (HACCP) methodology in certain developed regions. As both safety and trade regulations are harmonized at international levels, quantitative risk assessment and traceability will become integral components of aquaculture management. Developing countries have increased their share of the seafood export market to nearly 50 per cent of global trade, a significant portion being represented by aquaculture products (shrimp, salmon, molluscs, etc.), a percentage that should increase with the continued expansion of the sector.

The long-term viability of aquaculture development will be market driven, accounting for consumer demand and the capacity to adapt to the structure and legislative demands of the target markets. Important externalities will affect the production sector in achieving its goals, including the topics of sustainability, traceability and interactions with the environment. The development of schemes of best practice that incorporate the addressing of such issues, alongside quality schemes and safety management, will aid the production sector to achieve its goals. Discriminatory tariffs on trade should be avoided, particularly given the increasing importance of aquaculture as an export-earner and contributor to food security. The sectoral response to such developments must include a better understanding of the complex issues faced by the production and marketing of consumer food products within highly competitive markets, where the assumption of responsibility is essential, both for the product and the actions taken to produce it.

INTRODUCTION

At a global level, aquaculture is one of the fastest growing food production sectors (9.6 per cent/yr in the last decade), a fact that will ultimately change the way that fish is perceived as food. A key element of this observation is the change in the supply opportunities for fish and fish products from a wild source to a cultured one. Aquaculture product expansion has placed increased requirements on quality and food safety by consumers and regulators. There are few questions about the evident nutritional benefits of consuming fish but, by the end of the 1980s, developed countries arrived at the conclusion that classic fish (and food) inspection procedures, based on the analysis of samples of the final product and on generic hygiene measures, were not enough to provide the necessary level of protection to consumers. A preventive system called HACCP (hazard analysis critical control point) was adopted, and governments started to shift their regulations to HACCP-based systems. If one word has to be chosen to explain HACCP, it is "prevention."

An important point within the global market is the growing importance of international agreements that involve food (and fish) safety aspects. About 40 per cent of all fish produced are traded internationally, which means that there is a search for common criteria that facilitate or permit clear rules for compliance.

The tendency is, therefore, to move towards the harmonization of national regulations, meaning that such regulations could assure an equivalent level of food protection to consumers. This is a relatively easy concept to understand, but very difficult to implement and validate in practice. In turn, this increases the importance of internationally accepted guidelines, recommendations and standards, such as those of the Codex Alimentarius. The provisions related to food trade of the General Agreement on Tariffs and Trade (GATT) Agreement compound this tendency, and all of these aspects are interlinked.

The extent of regional and international trade in aquaculture products is difficult to analyse because trade in many aquaculture products is not yet well documented in all of the main producing countries. Aquaculture contributes primarily to domestic consumption but, at an international level, important trade has developed for a number of aquaculture products, and this has been one of the principal driving forçes for aquaculture development in many developing and developed countries.

INTERNATIONAL AGREEMENTS ON FOOD SAFETY

While there is no common agreement on food or fish safety at the international level, agreements do exist on food trade that have implications for food safety and quality matters. The important one regarding food safety is the Agreement on the Application of Sanitary and Phytosanitary and Measures (SPS) (GATT, 1994), which introduces two very important concepts. The first is that "to harmonise sanitary and phytosanitary measures on as wide a basis as possible, Members (countries) shall base their sanitary or phytosanitary measures on international standards, guidelines and recommendations, where they exist... (Article 3)".

The same agreement establishes that, regarding food safety, "international standards, guidelines and recommendations" to be taken as reference will be those "established by the Codex Alimentarius Commission (CAC) relating to food additives, veterinary drug and pesticide residues, contaminants, methods of analysis and sampling and guidelines of hygiene practice".

The second important point of the SPS Agreement is the adoption of the criterion that risk assessment shall be the basis for determining the appropriate level of both sanitary and phytosanitary protection. In this case, the SPS Agreement imposes a strong condition on national regulations that have not yet accommodated this criterion. Current HACCP-based regulations include a "hazard analysis" step, but risk assessment, which should be a part, is performed only qualitatively, and regulations do not refer to risk assessment explicitly. It is clear that the future development and evolution of regulations will be in this direction. All developed countries and a large number of developing countries have taken up regulatory HACCP-systems. These concern the safety of fish and fish products, which include products from aquaculture, and it is assessed that approximately 65 per cent of the total international fish trade is performed under such regulations. The greatest exception concerns the Japanese market, which accounts for about 24 per cent of the total fish market (demand), but which has no HACCP regulations as yet.

FISH SAFETY IN THE EUROPEAN COMMUNITY

The Member States of the European Community (EC) are subject to both European and national legislation, and considerable developments in laws affecting aquaculture have occurred in the last decade. These are separated into two clear sections that concern the final product and the aquaculture process.

The HACCP-based regulation for fish and fish products in the EC introduced the concept of "own health checks" (HACCP in terms of the EC regulation relative to fish products), and it is the Council Directive 91/493/EEC that laid down the health conditions for the production and the placing on the market of fishery products. EC aquaculture production follows Council Directive 91/67/EEC, which concerns conditions of stock health and health management and which, in practice, generated a "family" of regulations. For instance, Council Directives 93/54/EEC, 95/22/EC and 95/70/EC have successively amended Directive 91/67/EEC. This "family" of regulations is aimed basically at controlling fish diseases; they are only, in a second instance, complementary with the HACCP-based regulations aimed at preventing human diseases. The regulations related to aquaculture production do not imply accomplishment of the fish safety regulations, however. Although there are no contradictions between either regulation "family", their complementarity and coordination will surely increase with time.

FISH SAFETY IN THE UNITED STATES OF AMERICA

The basic regulation in the United States of America, which makes mandatory the implementation of the HACCP system in fish and fish products, is that of the United States Food and Drug Administration (FDA,5).

This regulation does not mention fish obtained from aquaculture specifically, but this source is included implicitly in the definition of "fish" ("where such animal life is intended for consumption") and in other paragraphs of the regulation (e.g., when "drug residues" is listed as one of the possible hazards to be accounted for). However, the regulation is specifically centred on processing, and does not apply to aquaculture production in itself.

As a general conclusion, one can say that such regulations are structured around a core principle that defines the HACCP-based system, and that this basic principle is complemented by a number of regulations that can be defined as "regulatory standards". Each species and product may be defined within a "family" of regulatory texts that outline the minimum requirements for each particular case or procedure.

Fish and fish products from aquaculture are included, either explicitly or implicitly, in such HACCP-based regulations. However, whereas the HACCP system is well defined for processing steps or procedures once the fish has been harvested, the application of the HACCP system to the entire aquaculture production chain is not so clearly established within the regulatory environment. Developments in this area can be anticipated.

Further evolution of the fish safety regulations may be expected with the introduction of new technical, scientific and operative knowledge and, in particular, regarding mandatory quantitative risk assessment.

All matters that relate to safety, quality and trade rely on the ability to identify and trace the product, but the traceability of food remains as an enormous problem to be solved. In its broadest sense, a traceability system would provide and allow access to all information concerning a product. In the case of aquaculture, this could mean following the history of a production batch from the final point of sale back to the hatchery. This would mean assuring knowledge of the steps encountered in distribution, packaging, processing and harvest, including the growing conditions, the feeds and therapeutic agents used, the broodstock parentage and any other relevant information.

Clearly, such a huge amount of information cannot be encoded physically with the product, but needs to be accessible via coding and linking to appropriate databases. The ability to trace and isolate a problem with a product is essential to any safety system, but the concept of traceability is much more than this, since it affords assurance, generates trust and, very importantly, can allow the use of important information for marketing and sales. Information is a new measure for adding value and will increase in importance.

AQUACULTURE AND TRADE

International seafood exports reached US$48 billion in 1998 (provisional FAO data), up from US$36 billion in 1990, but down slightly from the figures for 1996 and 1997. The share of developing countries in seafood exports grew from 43 per cent to 49 per cent between 1990 and 1997, giving net receipts of foreign exchange that rose from US$10.2 billion to US$15.8 billion.

The rapid growth in aquaculture production has made the sector important to the economy of many developing countries, and it has become either an important source of supply, or the main supplier, in the case of some products. For these farmed products, production fluctuations have had a significant impact on price trends. In general, however, aquaculture products have helped to stabilize traded supplies and to bring down prices over the years. Furthermore, several species that were once considered to be high-value luxury products have now become more abundant through aquaculture production, lowering prices and expanding markets. The extent of regional and international trade in aquaculture products is difficult to analyse because trade in many aquaculture products is not yet well documented in all of the main producing countries. Furthermore, because international trade statistics do not distinguish between wild and farmed origin, the exact breakdown between fisheries and aquaculture in international trade is open to interpretation.

This situation will change gradually, as aquaculture associations emerge in the main producing countries and start to keep records, and in response to various trade regulations that distinguish between farmed and fished origin (e.g., for shrimp).

TRADED AQUACULTURE PRODUCTS

In 1998, the main internationally traded products from aquaculture were shrimp, salmon molluscs and seaweed. Other species showing strong growth are tilapia, seabass and seabream.

Crustaceans

Marine shrimp is the most prominent product from aquaculture in international trade, and aquaculture has been the major force behind increased shrimp trading during the past seven to eight years. Shrimp is already the most traded seafood product internationally, with about 25 percent or 800 000 mt coming from aquaculture (FAO, 1999a).

Since the late 1980s, farmed shrimp has tended to act as a stabilizing factor for the shrimp industry. Therefore, the major crop failures in Asia and Latin America during the past few years (caused by disease problems) have had a significant impact on overall supply, demand, prices and consumption trends.

The major markets are Japan, the United States of America and, to a lesser extent, the EC, while the largest exporters of farmed shrimp are Thailand, Ecuador, Indonesia, India, Mexico, Bangladesh and Vietnam. Demand for shrimp is expected to increase in coming years, where Asian markets, such as China, the Republic of Korea, Thailand and Malaysia, will expand as local economies recover and consumers' demand more seafood. This trend is already reducing the availability of shrimp to traditional importers and will eventually put upward pressure on prices if supplies do not increase.

Trade in crab species has also increased with growing aquaculture production (1997: 165 000 mt). Especially important have been the exports of China (19,000 mt in 1998) to Hong Kong SAR China and Japan (INFOYU, 1999).

Finfish

In terms of total aquaculture output, finfish ranks first, with 49 per cent of the total production from aquaculture, of which the major part are carp species, which are consumed locally in the producing countries (mainly China and India). As opposed to shrimp, finfish aquaculture trade appears to be split between species having a high traditional demand and a "quality" image (e.g., salmon, seabass etc.) and convenience products (mainly fillets) of "cheaper" fish species (e.g., tilapia). The following species are the main products that are seen as being important in international trade:

Salmon: International trade in farmed salmon has increased from virtually nothing to more than 600 000 mt (1999) in less than a decade. The traded species are mainly Atlantic salmon and, to a much lesser extent, coho salmon, which accounted for 87 per cent and 12 per cent of 1997 salmon production, respectively (FAO, 1999b). The growth seen in trade has mirrored that of production, reflecting the fact that this activity is concentrated in a few countries that have limited domestic markets (Norway, Chile and the United Kingdom). Norway, whose main market is the EC, is the main exporter of Atlantic salmon. On the other hand, Chile, whose principal markets are Japan and the United States, is the main exporter of coho salmon and the second largest exporter of Atlantic salmon (FAO, 1999b; FAO, 1999c; Kontali Analyse, 1999).

As production volumes have increased, so has competition within the market and costs and prices have been driven down. At the current price levels (+/-US$4/kg CIF [Cost, Insurance and Freight) Europe), salmon has become a relatively medium-priced product in international seafood markets.

Trout: Trout is traded internationally at a much lower level than salmon, with exports reaching 135 000 mt in 1998 out of a total production of 463 000 mt. (FAO, 1999b). However, trout production is split, approximately equally, between portion-size production and large trout, and where 70 per cent of global production is in Europe. Consumption, particularly of portion-size fish, is concentrated in trout-producing countries, but Norway, Chile and Finland have been able to farm particular qualities of large-sized, heavily pigmented trout for the Japanese market, primarily as a replacement product for coho salmon. Japanese trout imports reached 60 000 mt in 1998 (Japanese Marine Products Importers Association, 1999).

Tilapia: Tilapia species have also shown a tremendous growth in output (production of 946 000 mt in 1997), and although international trade is limited, it is growing. This is observed especially between Latin America (Costa Rica, Ecuador and Colombia) and the United States, as well as between Asian producers (Taiwan Province of China, Indonesia and Thailand) and the United States and Japan. There is also modest trade between the EC and producers in Jamaica, Israel and Zimbabwe. The biggest exporter, Taiwan Province of China, supplies Japan with high-quality tilapia fillets for the sashimi market and ships frozen tilapia to the American market. Taiwan Province of China now exports about 35 per cent of its domestic tilapia production and supplies 80 per cent of the United States tilapia imports (1998). Thailand and Indonesia export less than five per cent of their production (Dey and Eknath, 1997). Vietnam has also recently entered the world tilapia market, and China exported 500 mt to the United States in 1998 (NMFS, 1999).

Tilapia has become the third highest imported aquaculture product, by weight, in the United States (1998 imports of 28 000 mt), after shrimp and salmon. United States imports rose 14 per cent by quantity in 1998 (NMFS, 1999). In the long term, tilapia prices are expected to decrease, and this should lead to greater exports to the United States, as well as to Europe, which is seen presently as being undeveloped as a market for tilapia.

Seabass and Seabream: In Europe, the marine fish farm industry of the Mediterranean has focused on the production of European seabass and gilthead seabream, and it intends to copy the success of salmon growers. Production reached almost 90,000 mt in 1999 (FEAP, 1999), of which nearly 90 per cent was exported from the country of origin, mainly to Italy and Spain. The principal exporter was Greece, which exported about 70 per cent of domestic production. Italy was, for a long time, almost the exclusive market for Greek production. However, as a result of market development efforts, some 15 per cent of Greek exports are now going to "new" markets (e.g., the United Kingdom, Germany, France etc.), and the share of these markets is expected to grow. It is not to be ignored that trade in live fingerlings is made from hatcheries in Italy, Spain and France to farms in Greece, Malta, Croatia and elsewhere.

As output of seabass and seabream has grown, market prices have more than halved between 1990 and 1999, dropping from US$16/kg to less than US$5/kg today. The rapid saturation of the market and the speed of the parallel decline in prices (down 50 per cent in five years, compared to the ten years for a similar drop in the case of Atlantic salmon) is attributed to several factors:

* the much smaller traditional market for these species (southern Europe), compared to Atlantic salmon;
* lack of diversified products;
* inadequate marketing and market development; and
* absence of technological advances (e.g., genetic improvement, efficient feeds and feeding strategies etc.) that could significantly improve productivity.

The substantial drop in prices should help to open new markets and expand existing ones, provided that acceptable profit margins can be sustained in production through improvements in productivity, diversification of products and intensified marketing efforts. American Catfish: American catfish is now the fifth most consumed fish species in the United States, measured at 0.4 kg/capita edible weight in 1998 (NMFS, 1999). Exports are limited, as production targets the domestic market, but some exports have started to Europe. The reason for the success of catfish is similar to that of tilapia: consumer demand for white, easy-to-prepare fillets.

Seaweed

Farmed seaweed production has increased in the last decade (6.8 million mt in 1997) and is now 87 per cent of total seaweed supplies. Most of the output is used domestically for food, but there is growing international trade. China, the major producer, has started exporting seaweed as food to the Republic of Korea and Japan. The Republic of Korea, in turn, exports some quantities of Porphyra (red seaweed) and Undaria (brown seaweed) to Japan (1998 exports of 26 000 mt). Chile produces over 100 000 mt, and is responsible for some 75 per cent of global production of Gracilaria sp., which is used for agar manufacture. Significant quantities of Eucheuma (red seaweed) are exported by the Philippines, Tanzania and Indonesia to the United States, Denmark and Japan. Total EC imports of seaweed in 1998 amounted to 58,000 mt, with the Philippines, Chile, Indonesia and Australia as the biggest suppliers.

Molluscs

International trade in molluscs is limited when compared to total output, with less than 10 per cent of production traded. The major importing markets are Japan, the United States and France, while major exporters are China and the Republic of Korea. The contribution of these farmed products to trade is uncertain.

Oysters, followed by clams, scallops and mussels, lead farmed mollusc production, but international mollusc trade concentrates on scallops and clams (fresh and frozen). Total fresh and frozen scallop imports have grown from 28 000 mt in 1985 to 65 000 mt in 1998, reaching US$510 million. Clam imports have risen from 33 000 mt to 152 000 mt in 1997, valued at US$257 million. Mussel imports have shown a downward trend after a peak of 175 000 mt in 1992, but have bounced back to 188 000 mt worth US$220 million. Oyster imports have grown from below 10 000 mt in 1985 to 28 000 mt (US$112 million) in 1998.

Live Seafood

The cultural preferences and growing affluence in Asia indicate a clearly positive long-term trend of live seafood commerce. The live seafood market is largely restricted to the restaurant trade and to consumers with a relatively high disposable income. Major market expansion is anticipated due to demand in China, but is also expected in Malaysia, Singapore and Taiwan Province of China, as well as in parts of North America with large Chinese communities. The potential for aquaculture to supply the market is promising. The sector is already supplying large amounts of shellfish and limited quantities of grouper, crabs and other species. Technological developments in the culture of preferred live food species will increase the contribution of aquaculture to supplies (Riepen, 1997).

Seed Supplies

There appears to be significant regional and international trade in the seed of cultured aquatic organisms, mainly from aquaculture sources, but this is poorly documented at present in most instances. Mention has been made above of the regional trade of fingerlings for Mediterranean seabass and seabream, but there is also international trade in wild glass eels (e.g., the recent development of large purchases of European eel elvers by China), "eyed" (fertilized) eggs of salmon and trout, the postlarvae of various cultured shrimps, Indian and Chinese carps, and others. There is also limited trade (in terms of quantity) in broodstock. Documentation of trade in seed and fingerlings will improve gradually as a response to concerns about the spread of pathogens and the movement of genetic material. One would anticipate that the concerns of traceability will also support such documentation.

ISSUES AFFECTING FUTURE TRADE IN AQUACULTURE PRODUCTS

The long-term viability and sustainability of aquaculture development, particularly in respect of commercial aquaculture, will be market driven, taking into account not only the consumer's requirements but also the structure and legislative demands of the target market, be it local or international. Some of the key issues that need to be addressed are mentioned in the following sections.

Externalities

Environmental and social concerns can influence markets for consumer goods and have already influenced farmed shrimp exports to North America and Europe in recent years. There is a growing desire for knowledge of what is being consumed, a position that, in some cases, is accompanied by accountability for consumption.

The importance of attaining sustainable aquaculture with no or limited negative externalities will force many exporting countries to adopt more sustainable production practices. The introduction of eco-labelling schemes will further accentuate this trend.

Where aquaculture is perceived as a non-traditional food-producing sector, it will have to further establish its credentials for sustainability, particularly when compared to fisheries and crop agriculture. This consideration requires to be extended to safety assessments, based on risk assessment and the precautionary approach, before entry into production of new or exotic species, including the potential use of the products from modern biotechnology.

Awareness of environmental and welfare issues is increasing, particularly in the developed countries, where purchase decisions can be influenced by adverse publicity or a lack of information. As livestock farmers, aquaculture producers are increasingly required to act responsibly and in line with standards expected of the activity. While such topics are partially addressed by the Food and Agriculture Organization's Code of Conduct for Responsible Fisheries (CCRF), it is increasingly recognized that standards applicable to international trade in aquaculture require harmonization.

At national levels, safety and quality management systems including Good Agricultural Practice (GAP) and Good Manufacturing Practice (GMP) should be developed and put in place in order to assure the production, distribution and sale of aquaculture products that are safe to consume and of high quality. Such measures require active and competent professional associations, working hand in hand with government, in order to be successful.

The collection, analysis and dissemination of relevant information should be facilitated in order to enable producers and industry operators to make informed decisions and to ensure consumer confidence in aquaculture products.

Quality and Safety

With growing concern about food safety, increasing efforts have been undertaken to improve the quality of food that is placed in the market, which evidently includes aquaculture. International codex standards cover aquaculture products, and the introduction of mandatory HACCP requirements for exports to the United States of America and the EC in 1997 has already had a great impact on trade in several aquaculture products.

Some countries have developed comprehensive HACCP plans for selected aquaculture products; for example, the United States now has plans for catfish, crayfish and molluscs.

In other countries, individual aquaculture producers have undertaken voluntary certification (ISO 9000) for control as well as marketing purposes. Such certification appears to be increasingly required for entry into markets such as multiple retail stores.

Alternative efforts include the development of industry-led quality schemes, which require government approval, to which individual producers can adhere. These schemes have controllers and strict operating procedures and conditions, serving to provide products of high quality and known origin.

In the field of HACCP, focus is increasingly on risk assessment by the operator, and this issue will put further institutional demands on exporting countries.

The effects of bovine spongiform encephalitis (BSE-mad cow disease) in cattle, the debate over genetically modified organisms (GMOs) and the increased awareness of toxin presence in the environment and food have contributed to a higher consciousness of the consumer to the quality and content of food in general, especially in the developed countries.

Actions to assure best practice, including traceability throughout the entire supply chain, are seen as being inevitable for assuring both the credibility and the sustainability of the aquaculture sector. This should not just apply to the aquaculture production process, but also include the feed content, particularly where additives and the potential use of GMO components are concerned.

SALES AND MARKETING

Attaining and maintaining consumer confidence requires considerable effort. It is no longer satisfactory to believe that production of a "high value" aquaculture product is a guarantee for the producer's long-term success.

Aquaculture history shows how technical success for rearing a "high-value" product incites production expansion, which leads to price drops or crashes. There are many different opinions as to the "whys" and "wherefores" of these circumstances, and market saturation is often given as the answer. There are, however, other explanations.

The long time required between investment in stocks and financial returns from sales can force producers to sell too early or at low prices, simply because of current cash needs, a circumstance that is punctual and rarely accompanied by marketing efforts. Such situations lead to market surpluses and price crashes in circumstances where better financial and production planning could assist.

The geographic dispersion of the activity and the relatively small amount of production per unit mean that there are a high number of sellers to markets that have few buyers (in number), a situation that gives keen competition and an advantage to the buyer. In addition, the aquaculture sector is relatively young and spends little money on marketing and promotion, particularly at the producer end of the scale.

Aquaculture produces perishable products with a short shelflife, particularly where products are sold fresh, meaning that distribution skills and production planning (to avoid surplus supplies to markets at specific times of the year) have to be honed to meet market demands.

The need for improved marketing and distribution skills is evident in many individual cases, where the grouping of producers under cooperative structures or as producer organizations is seen as being a response to such criticism.

The creation of efficient marketing systems, in which prices and costs are determined by supply and demand, moving to assure economic efficiency and sustainability, should be facilitated. The use of economic incentives and disincentives should be used to rectify market failure and the unsustainable use of resources.

DISTRIBUTION CHANNELS

The growing market share of multiple retail stores (super- and hypermarkets) in the distribution of foodstuffs has significantly changed patterns of production, supply and distribution.

For fish and fish products, these changes have had, in many markets, a profound impact on both the demand for products from aquaculture and the production sector itself. Modern distribution channels have developed buying criteria with precise requirements for quality, portions and sizes, price and delivery times that often can only be met by aquaculture producers.

This explains, to some extent, the success of products such as Atlantic salmon, which is prevalently sold through super- and hypermarkets in European markets. This has led to the virtual disappearance of the specialized fishmonger in certain developed countries and has imposed significant changes on the profession, in operating, marketing and organizational skills.

An example of this is the demand for packaged products, even for fresh, gutted fish. The demand now focuses on fixed weight packs (so that no individual weighing is required), particularly for processed products, using modified atmosphere packing (to assure freshness and a longer shelf-life) and where the sale price is attached prior to despatch to the supermarket. These conditions mean that, for the aquaculture producer to be capable of attaining this market, access to approved processing and packing facilities has become an essential part of working, as opposed to an option.

Labelling should follow the recommendations and codes of practice that are in line with the international requirements of the World Trade Organisation (WTO) and the Codex Alimentarius, demonstrating full traceability.

TARIFFS

Despite steady reductions in customs tariffs on both fish and aquaculture products in recent years, tariffs and import licenses continue to represent barriers to trade in many countries. This is especially the case in many fast-growing economies in Asia, but important markets such as Japan, the European Union and the United States of America all give competitive advantages to domestic producers of many species, especially in the case of processed products. On the other hand, producers in the developed markets argue that they are subject to higher levels of control and imposed regulatory investments that increase their production costs.

Average tariffs on imports from developing countries are now estimated at 4.8 per cent, a cut of 27 per cent from previous levels (FAO, 1995). The long-term trend, with growing membership in the WTO, will lead to further tariff reductions. The multilateral trade negotiations, which were to have started in late 1999, now show some delay in taking off, but could also have a large impact on future trade in fish and aquaculture products.

Measures that are taken to protect human life, animal life and health, the environment and the interests of consumers are increasingly seen as being potentially discriminatory, either in the nation of origin or in the exporting nation, depending on the point of view and the legislation in question.

FOOD SECURITY

Aquaculture is an important source of seafood, and the major part of the total output from aquaculture is consumed internally by the nations that produce, providing employment and an important nutritional contribution to society. Aquaculture has also become a significant source of foreign currency for many developing nations, since the products exported are usually the more valuable ones destined for markets in the developed world.

These revenues allow the countries to import other less costly protein and, as such, aquaculture plays an important role in food security, even where significant proportions of the output are exported.

CONCLUSIONS

The circumstances of producing and marketing fish and seafood produced from aquaculture are changing quickly. Technological advances have brought new species and higher productivity to the sector, which has developed to become an important contributor to national and international markets. Consumer demand for specific products, combined with good business opportunities, has contributed to the rapid development and restructuring of certain aquaculture subsectors, notably those concerned with export to developed countries. The industrialization of production and processing, alongside increased consumer awareness and legislative actions, has imposed quality and safety standards that are becoming requisite rather than optional.

Sustainability and environmental friendliness are also factors that are being linked to sectoral acceptability and, hence, influence quality, marketing and trade issues. The complex nature of the activity and its diversity mean that increased interregional and international cooperation should be encouraged, particularly in the fields of safety, quality and trade, where accent should be given to promoting a harmonized sectoral approach to the issues of concern.

3

Present Trends in Marine Fisheries

INTRODUCTION

Aquaculture used to be regarded as an infant when compared with crop and livestock farming and capture fisheries. In most parts of Asia, the activity has matured to become a clearly defined economic sector that is better organized and characterized by state patronage and strong private-sector participation. Many changes have occurred alongside the sector's development. Aspirations for higher yields through technological innovation have been tempered with concerns for sustainability. Economically, the drive for higher profits has been qualified by schemes to distribute benefits fairly. As a commodity industry, the purposes of producing more food, earning higher incomes and improving economies have been expanded to include ensuring food security, alleviating poverty, and promoting social harmony and prosperity.

These shifts in outlook, not confined to aquaculture and prompted by global and social forces, occurred in the last part of the previous century. It is in the midst of these changes that aquaculture development in Asia finds itself at the beginning of the new millennium.

STATUS OF ASIAN AQUACULTURE

Asia dominates global aquaculture production. This simple statement covers, however, the extreme diversity encountered within the sector, not only in the species reared and the technologies and farming systems employed, but also in the role and objectives of aquaculture development in different countries, the priorities placed by governments and the threats and constraints to its growth.

These considerations, in addition to the wide range of geographic locations, habitats and levels of national development, are important influences on how, and how fast, aquaculture will develop in the region. For these reasons, it would be better to look at the status and trends of Asian aquaculture by subregion, or on a national basis, with China classified by itself as a subregion in view of the sheer size of its aquaculture.

By the same token, India merits the same focus. Nevertheless, analysis of the overall regional status serves a useful purpose, particularly in trying to identify regional opportunities and actions.

Asia's contribution to total world production in 1997, following the International Standard Statistical Classification for Aquatic Animals And Plants (ISSCAAP) grouping, was: finfish-89 percent; crustaceans (marine)-80 percent; freshwater crustaceans-94 percent; molluscs-88 percent; aquatic plants-98 percent, and miscellaneous animals and products-99 percent.

The region provides 91 percent of global aquaculture production, where the top 10 Asian aquaculture producers are China, India, Japan, Republic of Korea, the Philippines, Indonesia, Thailand (also the top seven producers in the world), Bangladesh, Viet Nam and DPR Korea.

In 1997, the combined aquaculture production in Asia was 32.63 million mt valued at US$41.95 billion, an increase of 144 percent and 117 percent in weight and value, respectively compared to 1988. Aquaculture production in the region has been growing at a rate nearly five times faster than landings from capture fisheries, and its share of total fisheries landings in the region increased from 32 to 50 percent between 1988 and 1997.

PRODUCTION BY COMMODITY GROUPS

Finfish

Throughout Asia, the production of finfish has increased from 6.36 million mt in 1988 (having a value of US$10.1 billion) to 16.7 million mt (US$22 billion) in 1997, representing an average APR of 11 percent, while annual growth during the period 1992-1996 averaged 15 percent. In South Asia4, fish for food, mainly freshwater species, dominate production, with over 94 percent of total fish production in 1997. Freshwater and diadromous species occupy 92 percent of the subregion's total production, the remainder being crustaceans. Subregional production attained 2.2 million mt in 1997, worth US$22 billion.

In Southeast Asia, freshwater foodfish species comprise almost 30 percent of the total aquaculture output, mainly from Indonesia, Viet Nam, Thailand and the Philippines. While the overall production of diadromous species has been decreasing, milkfish contributed an additional 13 percent in 1995. Rising Indonesian aquaculture has been a major contributor to the increased production of freshwater species in this subregion. Aquaculture produced 1.5 million mt worth US$3.4 billion in 1997.

China is the major finfish producer in East Asia6, where total production was 13 million mt in 1997 worth US$16.1 billion. The production of freshwater finfish in China, mainly of carp and tilapia species, was over 12 million mt in 1997, accounting for 37 percent of regional aquaculture and 34 percent of global aquaculture. In recent years, diversification towards luxury freshwater species has been emphasized, including mandarin fish, freshwater crabs and prawns, eels and softshelled turtles. Nonetheless, carps accounted for 92 percent of freshwater production and represented 47 percent of national production in tonnage and 40 percent in value. Japan produces some 322 000 mt of finfish, worth US$2.8 billion, having an average value regularly higher than US$9/kg. This is mainly due to the production of significant quantities of Japanese amberjack (138 000 mt), silver seabream (80 000 mt), Japanese eel (24 000 mt) and rainbow trout (13 500 mt).

Around 14 percent of all finfish production in the world consists of carp, more than 90 percent of which is produced by aquaculture and where 80 percent of this amount is produced by China. In most countries, carp is consumed locally and, with few exceptions, producers of Chinese and major carp species have been unable to find markets outside Asia. Carps, as a group, are not traded globally. Carp supply in China, India and countries of the CIS [Commonwealth of Independent States], will probably continue to increase steadily, at least in the near future, in response to population growth and demand.

Tilapia has spread outside of Africa, its origin, and is now a common feature of aquaculture not only in Asia but also in Latin America and the Caribbean. Producers have access to domestic markets in several Asian countries, such as the Philippines, Thailand and China, and there is also an established and rapidly expanding market in the United States that is providing additional impetus to tilapia farming in several countries.

Small quantities of trout are farmed in Iran, Pakistan, the colder regions of India, and in Nepal, but Japan is the largest producer with some 13 500 mt produced in 1997. Trials are starting in Thailand to grow trout in the cooler climates (northern hills) as an income generating activity and largely for the domestic market.

The production of eels is significant in China, Japan and Taiwan Province of China, where the focal point for trade is Japan, which imported 11 000 mt of live eels and 45 500 mt of processed eels in 1996. Regional production has risen from 92 000 mt (1988) to 224 500 mt (1997). However, major growth has only occurred in China (>100 000 mt over the period), production dropping significantly in Japan (from 40 000 mt to 24 000 mt) and in Taiwan Province of China (52 000 mt to 22 000 mt). Furthermore, the 1998 value of the eel output (US$314 million) in Japan was a huge drop from the 1995 peak of US$577 million, because of the decrease in quantity produced (from 29 131 mt in 1995 to 21 971 mt in 1998) and average price of the product from (US$20/kg to less than US$15/kg.

As production capacity has grown, disease and a shortage of elvers of Japanese eel (Anguilla japonica) have caused seed supply problems in the region and initiated the trade of elvers of the European eel (A. anguilla) from Europe to Asia, particularly, China. Recently, a resurgence in the recruitment of the Japanese elver has alleviated the situation.

Crustaceans

Aquaculture of crustaceans, mainly shrimp and crabs, yielded 539 000 mt in 1988, worth US$3.5 billion. By 1997, this had risen to 1.1 million mt (APR 9 percent), having a value of US$7.2 billion (APR 9 percent).

Shrimp production in 1997 was worth US$6.9 billion at farm gate prices to the region, representing a total of 902 000 mt. The value added through processing is also very significant, making it worth more than US$6.9 billion to the region. Penaeid shrimp is the main crustacean aquaculture product that is traded internationally but, although this is an important source of hard currency in many developing economies, shrimp are much less significant as a food source for domestic consumption. The total volume of cultured penaeid shrimp is now close to half that produced by capture fisheries. In volume terms, the rate of increase in production of penaeid shrimp is tapering off in Asia, but one can see a significant increase in the production of freshwater crustacean species, Macrobrachium rosenbergii, mainly in Bangladesh and China.

The growth of penaeid shrimp production has been predicted to slow down in the immediate future. One of the main reasons for this short-term prediction is the slower rate of economic growth in Japan, the world's largest market for shrimp, as well as the slower economic growth expected in the other developed economies. In addition, the management of shrimp culture is not at a uniformly high level. Furthermore, the emergence of stringent environmental regulations may slow down expansion in Asia.

In South Asia, crustacean culture is dominated by penaeid shrimp, which increased from 37 000 mt in 1988 to 134 000 mt in 1997 as a result of increased production in India, Bangladesh and Sri Lanka valued at US$858 million. Growth slowed in 1997 due to disease outbreaks and India's sanctions on shrimp culture. Of note is the large harvest reported in 1997 for the freshwater giant river prawn (M. rosenbergii) in Bangladesh, 48 000 mt valued at US$497 million.

In Southeast Asia, the production of shrimp and prawns grew from 220 000 mt in 1988 to 562 500 mt in 1997, where Penaeus monodon (giant tiger shrimp) represented 450 000 mt in 1997 (80 percent). The decade reviewed (1988-1997) shows the phenomenal growth of this sector in Thailand (+242 percent), the world's leading producer of this species, Viet Nam (+293 percent) and Indonesia (+115 percent). After production stagnation during the late 1990s, Thailand has rebounded with production of an estimated 250 000 mt in 2000.

East Asia has witnessed significant reductions in shrimp production, where production of the fleshy prawn (Penaeus chinensis) dropped from a high of 207 000 mt in 1992 to a low of 64 000 mt in 1994, recovering to 146 000 mt in 1997. Taiwan Province of China's giant tiger shrimp production dropped from a peak of 78 500 mt in 1987 to only 5 000 mt in 1997. Overall, the subregion's production of shrimp and prawns has decreased from 246 000 mt to 163 000 mt, but has shown signs of recovery in recent years. Other crustaceans for which production is increasing are crabs, notably in China, and after a period of decline, production of freshwater prawn is also rising with increasing outputs from China and Bangladesh. In West Asia, crustacean production is nominal and restricted to penaeid shrimp. In Iran it is small but increasing.

Molluscs

Mollusc aquaculture is dominated by East Asia (98 percent of the region's production), the remainder being attributed to Southeast Asia. Regional production reached 7.6 million mt in 1997, valued at US$7.6 billion, compared to the 1988 production of 2.6 million mt worth US$2.5 billion, an APR of 13 percent in volume and 14 percent in value.

Molluscs are almost always reared for sale, rather than household needs, usually being sold to markets located near the place of production. Mollusc culture in Asia was not affected much by the 1997-1999 economic downturn and will continue as a source of growth in aquaculture production. Cockles and mussels are promoted as cheap sources of protein, while oysters tend to be more of a luxury item, consumed domestically and in some countries as an export item Pearl oysters are significant industries in China, Indonesia, Japan and the Philippines and to a lesser extent in India and Thailand.

Big increases in production have occurred throughout the region during the period examined, but most significantly for:

* Japanese carpet shell (China: from 186 000 mt to 1 257 500 mt);
* razor clams nei (not elsewhere included) (China: from 138 500 mt to 354 000 mt);
* Pacific cupped oyster (China: from 447 500 mt to 2 328 500 mt); and
* Yesso scallop (China: from 129 500 mt to 1 001 500 mt; Japan: from 180 000 mt to 254 000 mt)

Elsewhere in the region, oyster production has stagnated. Mussel production, in general, has declined everywhere, with the exception of the increases reported for the Korean mussel in the Republic of Korea (8 000 mt to 64 000 mt).

Aquatic Plants

East Asia and Southeast Asia produce the bulk of aquatic plants within the region, where seaweeds, mostly brown or red, dominate the volumes reported.

These subregions provide almost all of the brown, red and green seaweeds produced in global aquaculture. While aquatic plant production almost doubled between 1988 and 1997 in East Asia, rising from 3.5 to 6.3 million mt, the contribution to aquaculture production dropped from 32 percent to 23 percent. In Southeast Asia, the pattern is different, production growing over the same period from 0.34 to 0.76 million mt, the contribution changing from 22 percent to 25 percent. The region's total production was 7 million mt in 1997, valued at US$4.8 billion, figures that indicate APRs of 8 percent and 5 percent, respectively. The average value/kg for this group has eroded from US$0.84 to 0.68, an APR of-0.2 percent.

The highest reported global production of any cultured aquatic organism for 1997 was the kelp Laminaria japonica, totalling over 4.4 million mt. In the top 10 species, carps have consistently led production, followed by plants and shellfish (1988); by 1997, shellfish had overtaken aquatic plants in this position. Comparison of the top ten aquaculture species in the Asian Region, ranked by production importance in 1988 and 1997 is given.

One can note that the top ten species contribute two-thirds of the region's production, but that carps have displaced aquatic plants from its dominant position and shellfish production, having tripled in volume, now occupies second place in the groups containing the top species.

Most notable is the rise to the top, from 5th place, of the giant tiger shrimp (Penaeus monodon), which contributed 9 percent of the total value of Asian aquaculture in 1997, although it is 17th in the production ranking. Thailand contributed more than half this figure, despite disease problems. The production of P. chinensis, which is predominantly cultured in China, was ranked 42nd. While the production and contribution of the major carp species have increased significantly, the average US$ value/kg has gone down; for the examples cited in this analysis, reductions of 12-30 percent have been seen.

Most of the shrimp species have maintained or increased their US$ value/kg. During the same period, the value of the Japanese eel has halved, while certain fish species (milkfish, rohu, Japanese amberjack and tilapias) have increased.

ENVIRONMENT

If one separates aquaculture production by environment, the proportion of freshwater species in total Asian production increased from 43 percent to 49 percent between 1988 and

1997, an average APR of 12 percent. Brackishwater production had an average yearly growth of 5 percent and marine production 10 percent.

The proportion of brackishwater aquaculture decreased from 7 percent to 4 percent while experiencing growth at an APR of 10 percent during the same period. Similarly, marine production had average growth of 10 percent, but its share decreased from 50 percent to 47 percent.

Freshwater finfish, in particular the Chinese and major carps, accounted for the greatest share (48 percent) of regional aquaculture production in 1997, while aquatic plants contributed 22 percent. A key factor in the rapid rise in the production of some of the species cited has been the increased availability of hatchery-produced seed.

MAIN CULTURE SYSTEMS

Inland fresh-and brackishwater ponds, tanks and reservoirs comprise the largest production areas in Asia, with some 1.99 million ha in China. Similarly, there are 147 000 ha in Bangladesh, 900 000 ha in India, 462 000 ha in Indonesia, 255 000 ha in the Philippines, 320 000 ha in Thailand and 392 000 ha in Viet Nam. Viet Nam, Indonesia and the Philippines have more brackishwater than freshwater ponds and which are devoted to shrimp culture as well as other species, such as milkfish in Indonesia and the Philippines.

Freshwater ponds in Viet Nam, Cambodia, Laos, Bangladesh and India are generally small and used to grow food fish that are low on the trophic level. These are usually integrated with other operating systems, such as the 127 000 ha of VACs (vegetable-fish-livestock plots or gardens) in Viet Nam, and the undrainable composite-culture ponds in India.

The Network of Aquaculture Centres in Asia-Pacific/Asian Development Bank Study (ADB/NACA, 1998) found that almost all carp farming in the region is a polyculture of indigenous and (often) exotic carp species, with the exceptions of Korea and Indonesia, where monoculture of common carp was common. The carp farming systems also commonly contained other non-carp species. There are three distinct areas of carp polyculture in Asia: South Asia has the major carps; China and cooler waters use Chinese carps, and Southeast Asia uses a mixture of exotic and indigenous carps. For example, in Cambodia, Laos and Thailand, indigenous cyprinids such as Puntius spp. are more common. Thailand's freshwater ponds are operated for a mixture of objectives. The products of school and village ponds provide food and income among poor families, and these comprise nearly 13 000 units, ranging in size from one-third of a hectare (school ponds) to 8 ha (community ponds). All in all, Thailand has a total of 247 300 ha of inland aquaculture ponds, comprising 122 000 farms and producing annually 196 000 mt of mostly food fish. Commercial freshwater ponds are used to raise higher-priced indigenous species, such as catfish, snakehead, freshwater prawn and tilapia. Many of the commercial entities are integrated farms.

Cambodia has the highest percentage (97 percent) of integrated farms within the semi-intensive carp production activity, perhaps a result of the introduction of this system by nongovernmental organizations (NGOs) as opposed to tradition. Thailand follows with 62 percent of farms integrated, primarily with chicken farming and tree crops.

Brackishwater ponds are generally devoted to higher-value species, shrimp being the most popular, followed by other crustaceans and diadromous species. Indonesia has extensive brackishwater ponds (385 000 ha] used for shrimp and milkfish. The Philippines has more than 250 000 ha producing milkfish and shrimp; lately, trials have been made with salt-tolerant tilapia being rotated or polycultured with shrimp to determine its effects in controlling shrimp diseases. In India and Bangladesh, there is a tradition of extensive culture of shrimp in large water bodies.

The potential for integration of rice and aquaculture has long been recognized. China has shown that rice-fish systems are indeed productive and, by also using the system to nurture fingerlings, can generate additional income for rice farmers over a short period. China has 1.3 million ha applied to the paddy cum fish system and an additional 281 000 ha of paddy used for fingerling growing. China's production from the rice-fish system was 455 000 mt in 1997 or almost 500 kg fish/ha.

Freshwater fish culture integrated with rice is also found in South Asia and Southeast Asia. Alternate cropping of rice and shrimp-a crop of rice in the wet season followed by a crop of shrimp in the dry season-is also found in some seasonally brackishwater deltas in Southeast Asia (notably Viet Nam) and South Asia (India and Bangladesh).

China's freshwater fishponds yield an average of nearly 5 mt/ha (9 million mt of fish from 1.9 million ha). Culture-based fisheries in open waters are likewise very productive; 880 000 ha of freshwater lakes yielded around 810 000 mt of fish in 1997, and 1.56 million ha of reservoirs provided 1.16 million mt. China has also exploited the potential of canals, producing 602 000 mt of fish from 370 000 ha of these waterways. The system used within canals usually involves cages or pens.

The NACA/ADB study on aquaculture sustainability and the environment (NACA/ADB, 1998) found that only 24 percent of surveyed freshwater farms in China were integrated, reflecting farm intensification and perhaps, changes in traditional farming practices. The report cautions that this observation needs further confirmation. China's mariculture (in shallow coastal waters and bays) is likewise characterized by the integration of various species (e.g. seaweed, mollusc an cucumber within a column of coastal water) for an intensive use of the marine environment.

In 1997, China harvested almost 4.7 million mt of seaweeds, 6.5 million mt of molluscs (including scallops and abalones) and 21 500 mt of other coastal aquaculture species. The areas where mariculture is done were made up of 7.9 million ha, including shallow coasts (3.7 million ha), bays (0.5 million ha) and mudflats (3.7 million ha).

Culture-based fisheries in open-water systems, such as oxbow lakes, are significant in India and Bangladesh, where production yields are around 500 kg/ha for native and exotic carps. Stocking involves both native and exotic carps, although some recent stocking programmes in Bangladesh are giving more emphasis to indigenous carps due to concerns over the potential impacts of stocking of exotics on native fisheries and aquatic biodiversity.

PRODUCTION LOSSES

The first attempt, at a regional level, to quantify the effects of disease occurrence was made by NACA (ADB/NACA, 1991) in 1990. At that time, losses incurred from disease outbreaks were estimated to have been worth US$1.4 billion a year in the developing countries of Asia, and these have increased since. Reports from China suggested that in 1993 alone, shrimp diseases caused losses of US$1 billion. An ADB/NACA farm survey in 1994-1995 (ADB/NACA, 1998) indicated that for carp and shrimp farming, losses from diseases and environment-related problems were valued at around US$300 million a year. Furthermore, external influences that are associated with disease problems (including chemical use, environmental impacts, trade issues etc.) have yet to be properly assessed. Generally, economic losses are likely to increase in scale as aquaculture expands and intensifies, affecting both large-scale and small-scale production units.

PRODUCTION TRENDS AND GROWTH RATES

A noteworthy, but perhaps disturbing trend seen in the statistics for the Asian countries/territories from 1988 to 1997 is that the average aquaculture growth (APR) has been 10.4 percent by weight and 9.0 percent by value, but production has not shown steady growth, with year on year increases varying between 1.6 percent in 1990 and 18.6 percent in 1992. After peaking at 18-19 percent in 1992-1993, the rate of increase has slowed down steadily to 5.7 percent in 1997.

This observation coincides primarily with the trends seen for Chinese aquaculture. Some countries within the region have shown production decreases, with a steady decline of production through the period. These are Japan, China, Hong Kong8, Democratic Republic of Korea and Taiwan Province of China.

A striking trend in Japanese aquaculture has been the decline in production of the major species. For instance, the yields of Japanese amberjack have declined from a high of nearly 170 000 mt in 1995 to 147 000 mt in 1998, while laver dropped from a peak of 483 000 mt (1994) to under 400 000 mt (1998). Pacific cupped oyster production has also shown a steady decline. Of the major mariculture species, only silver seabream and Yesso scallop have shown trends towards rising production.

The annual rate of production growth has been consistently higher in China than in the rest of the region, reflected by the increased rate of difference in the compared production volumes. If aquaculture values are compared, the picture is slightly different. This figure shows that the difference in value is proportionately much less than that of production, confirming the effects of diversification and the maintenance of value in aquaculture in the rest of Asia. Indeed, in China the APR for value has grown and shown stability from 1990-1996, dropping off in 1997. In the rest of Asia, the value APR grew steadily from 1989, but has since declined regularly from 1994 onwards, even showing negative figures in 1996 and 1997. Theses circumstances meant that Chinese production alone had a total value higher than the rest of Asia from 1996.

Demand for Aquatic Produce

The projections for the average annual population growth rates in the region from 1998 to 2020 are:

* 0.8 percent for East Asia, growing to 1.66 billion;
* 1.6 percent for Southeast Asia, growing to 655 million;
* 1.7 percent for South Asia, increasing to 1.72 billion; and
* Iran will have 89 million and Australia 22 million.

The total equals 4 124 million people or two-thirds of the earth's population. Just to maintain, region-wise, the supply of 17.2 kg of fish per capita (1995 average), the Asian Region would require more than 70 million mt of food fish in 2020, to be supplied from capture fisheries and aquaculture. A look at the situation by region would show differences, but all the same, the indications are for aquaculture to increasingly fill the fish supply requirements of the populations.

In China, aquaculture already contributes more than 60 percent of the total fisheries production and, in 1995, the per capita supply of fish was 22.2 kg, implying a need for 32.2 million mt by 2020 assuming a constant per capita fish supply. If the same supply proportion were to be maintained, 19 million mt would be needed from aquaculture. Southeast Asia's average per capita supply is now 24.7 kg a year; with a population of 655 million it will require 16.2 million mt of food fish. In assuming that 30 percent is to be contributed by aquaculture, almost 5 million mt of food fish need to be supplied by aquaculture.

South Asia will need a total of 8.1 million mt of food fish. In 1997, only 2.33 million mt were produced by aquaculture, 97 percent of this being food fish. If the same ratio is maintained, the South Asian Region will need 7.8 million mt by 2020.

With a total requirement of 35.7 million mt of cultured food fish forecast for 2020, the concern for supply is evident. Since the wild catch is stagnating and forecast to slow even further, aquaculture will have to increase its contribution beyond its current levels.

The Food and Agriculture Organization of the United Nations (FAO) predicts that global aquaculture is likely to show sustained growth in the medium term and could attain 35 to 40 million mt of finfish, crustaceans and molluscs by the year 2010 (FAO, 1998). What this analysis clearly points out is that there is now a burden on aquaculture to meet the future demand for aquatic products, especially that the continued capacity to sustain further productivity increases in all the food production sectors is facing the following threats (UNDP, 1998): losses in genetic resources and biodiversity (e.g. 4 percent of fish and reptile species are threatened with extinction); loss of aquatic habitat such as coral reefs and mangroves, and severe depletion of commercial fish stocks (25 percent of fish stocks for which data are available are either depleted or in danger of depletion, and another 44 percent of fish stocks are being fished to their biological limit).

Aquatic resources are not unique in facing such degradation, as 9 million ha of land are extremely degraded, with their original biotic functions fully destroyed, and 10 percent of the earth's surface is at least moderately degraded, and the availability of safe and clean water has declined steeply to 60 percent (in 1996) of its 1970 value (UNDP, 1998). These trends, which show no signs of changing, indicate that future production of food from aquaculture will face significant challenges in access to and use of resources.

Another crucial point that should be emphasized is that the countries in this region have the highest population densities in the world. There are many issues that aquaculture has to face, but the fundamental one for Asia is that of supply and demand, which has been illustrated above. The critical concerns identified for resolution of this issue relate to social, economic, biological, environmental and institutional questions, which must be resolved with a minimum of conflict and most efficient use of resources. To put this into a positive light, aquaculture must be developed in a spirit of social harmony, environmental sustainability and economic progress.

TRENDS AND ISSUES

International Trade in Aquaculture Products

Information on the trade in aquaculture products is rarely separated from trade in fisheries products, making it difficult to analyse trends in the trade of aquaculture products. Comments here are therefore based mainly on fisheries products (FAO/RAP, 2000). By volume, Asia is a net importer of fishery products, mainly because of Japan's trading patterns where an annual trade deficit (exports less imports) of more than 3 million mt has been registered since 1994. If Japan is excluded from the trade data, the developing countries of Asia would register a surplus of only 35 000 mt a year in the triennium up to 19979. The major net importers (with volumes over 100 000 mt) of fish are China, followed by the Philippines and

Malaysia. The major exporters with volumes greater than 100 000 mt are India, Indonesia and Thailand. Of these, aquaculture production probably makes the biggest contribution in Thailand.

In terms of financial value, the countries with developing economies are responsible for more than half of the region's fishery exports, and there is a dramatic change in the pattern of imports and exports. Exports from developing countries in Asia are mainly high-value products that are directed primarily to developed economies. In the triennium ending in 1997, Thailand was the leading exporter in the region, exporting US$2.92 billion annually, a significant portion of which was cultured shrimp. Other major exporters with more than US$1 billion of fish products per year over the triennium include China, India and Indonesia. In terms of financial value, the countries with developing economies are responsible for more than half of the region's fishery export trade, which is directed primarily to developed economies. Between 1993 and 1996, Thailand was the leading world exporter of fish products, at annual values of around US$3.4 billion (although displaced by Norway's fishery exports in 1997). Japan is the leading importer of fish products in the world, with a value of US$15.5 billion in 1997. United States absorbs about 10 percent of the total value and these two countries and the European Community (EC) (including the value of the intra-EC trade) import 75 percent (in value) of internationally traded fishery products.

Shrimp Trade

In financial value, shrimp is the most important traded aquaculture commodity in Asia, and Thailand is the world's main supplier of cultured shrimp, with an output of over 200 000 mt. In 1997, the shortage of shrimp on the world market was acute; but Asia, as a whole, was able to maintain its share of the United States market, and smaller exporting countries in the region such as Indonesia and China performed well, although China has of late rarely exported any shrimp.

Indian shrimp exports to Japan increased by 6.6 percent to a record 59 100 mt, while total exports in the late 1990s have exceeded 100 000 mt, of which around half were cultured.

The reasons for increased import to Japan included EU restrictions on Indian seafood, which started in August 1997.

In 1997, Japanese imports of shrimp fell by 7 percent to 267 200 mt, the lowest figure in nine years, reflecting a downward trend that was apparent throughout 1997. The United States shrimp market was very strong, owing to the country's expanding economy and to the high value of the US dollar, and prices rose by 20 percent in one year. United States imports expanded by 10 percent in 1997, overtaking Japan as the world's major shrimp market for the first time. Within the region, for markets such as China (including Hong Kong), Singapore and Malaysia, the imports of fish and fishery products for domestic consumption continued to be lower in 1997 than in 1996 due to the economic problems experienced.

Seaweed Trade

Seaweed exports bring significant earnings to the Philippines, China, Indonesia and the Republic of Korea. The Philippines produced over 620 000 mt in 1997 of mostly Eucheuma seaweed, from which the colloid carageenan, which is the export product, is extracted.In East Asia, seaweed (Laminaria and Undaria) are also produced and particularly Undaria, exported.

Mollusc Trade

Mollusc exports from the region, including intra-regional trade, increased from 240 000 mt in 1988 to 335000 mt in 1997, peaking at 380 000 mt in 1996. Important exporting countries are China, Democratic Peoples Republic of Korea, Republic of Korea, Malaysia and Thailand. The predominant products are clams, scallops and oysters.

Finfish Trade

The trade in live marine fish, particularly reef fishes, has been growing steadily, spurring work on their aquaculture. Market analysis done in 1995 indicated that the total seafood market in Hong Kong and southern China, the main markets at present, was over 220 000 mt a year, within which the estimated annual demand for high-quality live reef fish was 1 600 to 1 700 mt. This study forecast compound growth rates of more than 12 percent, in other words, doubling every six years.

In 1997, Hong Kong consumed 28 000 mt of live fish, of which groupers represent 35 percent by weight and 50 percent by value. Tilapia, filleted or chilled, is a growing export species. Taiwan Province of China supplied half of the 50 000 mt of tilapia consumed by the United States in 1998. Other exporters of tilapia to the United States market, but as yet in small quantities, are Indonesia and Thailand.

Singapore is also a significant market for high-value marine fish, although local cage farming has mainly satisfied its domestic consumption. To meet expected growth in demand, it has initiated the move to farm marine fish in offshore deep-water cages, targeting annual production of 40 000 mt that would be supplied with fry from local hatcheries.

The breeding and growing of ornamental fish and aquatic plants is also a growing industry. The value of world trade (wholesale and retail) in these commodities has been estimated at US$5 billion, generating employment and income to breeders and growers. Singapore, Malaysia and Thailand have well established and mature industries. Countries such as Sri Lanka and the Philippines have discovered this sector fairly recently, seeing that the activity offers opportunities for rural families to earn income from cultivating freshwater ornamental fishes.

QUALITY OF AQUACULTURE PRODUCTS

The importance of aquaculture production in Asia and in particular that of freshwater fish, most of which are sold in the domestic market, has a significant bearing on the concerns of human health associated with products from aquaculture. The most common method of producing low-value freshwater fish in developing countries, especially in low-income food deficit countries (LIFDCs), is to use a combination of fertilizer application and supplementary feed. Polyculture of species that have complementary feeding habits is a general practice, especially with major and Chinese carps. However, the patterns of production are changing rapidly, with a large increase in more semi-intensive farming.

There will be further emphasis on food safety issues in the future. In 1997, a WHO (World Health Organization)/FAO/ NACA Study Group (WHO/FAO/NACA, 1999) looked into the food safety issues associated with aquaculture and concluded that there were considerable needs for information, recognizing that such knowledge gaps could hinder the development process. Education is needed to increase the awareness of potential hazards in aquaculture and their management. The Study Group also recognized the difficulties in applying Hazard Analysis Critical Control Point (HACCP) procedures to small-scale farming systems and that food safety hazards associated with aquaculture vary by region, habitat and environmental conditions, as well as by methods of production and management.

In Asian shrimp production and processing, food safety is given increasing attention, largely because of the quality standards required by importing countries but also because of the higher profile given to environmentally friendly practices in the aquaculture sector.

AGREEMENTS REGULATING INTERNATIONAL FISH TRADE

International rules and regulations play a major role in governing the trade of fishery, affecting aquaculture products in both developed and developing economies. Several major international agreements and codes of conduct include sections that relate to aquatic animals and their products.

In the Asia-Pacific Region, however, there is a wide difference in knowledge and expertise between different countries Furthermore, some countries in the region are members of organizations such as the Office International des Épizooties (OIE) and the World Trade Organization (WTO), and comply with the terms of such agreements, while others who are not signatories are obviously not bound by the terms of such agreements.

Regulatory and Management Frameworks for Aquaculture

Agencies Responsible for Aquaculture Development

In Asia, the primary responsibility for aquaculture production usually resides within one government ministry or department. By contrast, the management of the resources upon which aquaculture depends, particularly water and land, is normally the responsibility of other ministerial departments. This situation inevitably causes the potential for conflicts between different departments, and confusion or lack of clarity in the application of policy.

The increasing importance of aquaculture strongly argues for governments to give priority to developing clear, well-formulated, implementable and realistic national and local policies for aquaculture development, based on financial, social and environmental sustainability. As the private sector is the key to successful and sustainable aquaculture development, the views of industry should be taken into account in respect of policy formulation, research and development.

Participation

The adoption of a "participatory approach" is receiving increasing attention in both aquaculture planning and management. Experience with coastal fisheries management has shown, in general, that the failure to include coastal residents in natural resource management can lead to a lack of community compliance, resulting in resource depletion and conflicts.

This situation can be particularly acute when the government's capacity to enforce laws and regulations is limited.

A promising approach to this problem is "co-management", which involves the cooperation of the local community in both establishing and enforcing local management rules, with the requisite support from government. The co-management approach has proved useful for community-based management of some coastal capture-fishery resources, but the devolution of ownership and management of resources towards local people and communities should be explored for coastal aquaculture. The need to develop effective local arrangements for management of aquatic resources will also be stimulated further in some countries by the trend towards devolution of decision-making powers to local government bodies, as in the Philippines, Thailand and Viet Nam.

It is also apparent that the views of non-aquaculture groups need to be considered when formulating policy and implementing aquaculture development; without such views and contributions, some problems relating to conflicts of interest may be too difficult to fully resolve. Such issues are quite new, but there is evidence of increasing involvement of various aquaculture interest groups, including the private sector, local farmers and others, in aquaculture policy formulation. In some countries, the government and farmers have entered into discussion with NGOs and non-aquaculture groups for the formulation of management strategies. Examples have been seen in Sri Lanka, where NGOs have participated in aquaculture policy discussions, and in Malaysia, where input from various government, farming, academic and NGO interest groups was provided in the formulation of the shrimp and marine fish "Codes of Conduct".

Legislation

In many countries, aquaculture legislation is poorly developed and frequently consists of a few articles contained within a law regulating capture fisheries. During the last few years, interest has grown to develop a comprehensive regulatory framework for aquaculture whose goals would be to protect the industry, the environment, other resource users and the consumer, and to clarify rules for resource use. This consideration is driven by a variety of factors that include the following:

* greater political attention to the sector as its economic importance and potential become more apparent;
* realisation that inappropriate laws and institutional arrangements can severely constrain the sector's development;

* evidence of environmental damage and social disruption as a result of the rapid and largely unregulated expansion of shrimp farming; and
* a growing emphasis on examining production methods as a means of improving the quality and safety of aquaculture products.

While capture fisheries are generally regulated by a single government department or ministry, aquaculture is frequently regulated by many agencies under a variety of laws. In this circumstance, the formulation of a comprehensive regulatory framework for aquaculture is legally and institutionally complex. Typically, the tasks involve drafting or amending legislation that addresses a variety of issues such as land use planning and tenure, water quality and environmental issues, fish movement and disease, pharmaceutical and chemical use, food quality and public health.

It also requires clarification of responsible authorities and the establishment of the arrangements to ensure cooperation and coordination between many different institutions that have jurisdiction over the item or issue concerned.

INTERNATIONAL DEVELOPMENTS

The adoption of the Code of Conduct for Responsible Fisheries (CCRF) by the 1995 FAO Conference (FAO, 1995) promises to provide a significant influence on the development of regulatory systems for aquaculture in the coming years. Article 9 of the Code deals with aquaculture development and sets out a wide range of relevant principles and criteria. In addition, the "Jakarta Mandate", which was adopted by the second Conference of the Parties to the Convention on Biological Diversity in 1995, provides useful guidance regarding aquatic biodiversity and related environmental aspects that should be taken into account when developing coastal aquaculture.

During 1994 and 1995, several regional workshops10 were organized to examine environmental, legal and institutional issues associated with aquaculture. Recently, FAO published a compendium on aquaculture and inland fisheries legislation in the Asian Region (FAO, 1996). Although new national laws to regulate aquaculture comprehensively may be desirable in many countries in the region, other options are now being explored. This is because the time required to develop and pass new comprehensive legislation is likely to be several years, while the rapid development of the sector has created an urgent need for regulation. These options include the enactment of regulations that exist under existing legislation and the application of voluntary approaches such as guidelines and codes of practice.

Creating incentives to encourage compliance or disincentives for noncompliance or failure to join the scheme can enhance the effectiveness of voluntary codes of practice. One example is the policy that was adopted in India in 1995 by the National Bank for Agriculture and Rural Development (NABARD), which was a major source of refinancing for brackishwater prawn farming in the states of Andhra Pradesh and Tamil Nadu.

This policy sets out the conditions to be met before NABARD will provide credit to banks for aquaculture developments. Similarly, Malaysia is considering the requirement that aquaculture enterprises which are given preferential "pioneer" tax status must adhere to a code in order to retain the various tax incentives accorded. Clearly, there is scope for increasing the use of such conditional public and private-sector funding to encourage the application of sustainable aquaculture practices.

The Australian State of Tasmania has recently passed legislation that provides another interesting example of what can be done to address this issue. The new laws (notably the Marine Farming Planning Act [1995] and the Living Marine Resources Act [1995]) impose that the preparation of marine farming development plans must cover areas, rather than sites, and provide for broad community participation in the preparation of these plans. An environmental impact assessment must also be done, with the establishment of a marine farming zone, before site leases are granted to marine farms.

Malaysia is developing an integrated regulatory system for aquaculture, in the short to medium term, but without formulating a specific aquaculture act. Instead, new regulations will be incorporated under the existing Fisheries Act, and a voluntary Code of Responsible Aquaculture Practices will be introduced for cage culture and shrimp farming. These measures will be supported by incentives, and institutional structures will be strengthened to ensure the effective monitoring and continuing formulation of aquaculture policy. Many regulatory systems could be improved by including mechanisms that are designed not only to prevent or reduce the risk of environmental harm but also to help right any damage that may occur. The Tamil Nadu Aquaculture (Regulation) Act of 1995 not only set out conditions to improve the siting and management of aquaculture facilities, but also established an "Eco-Restoration Fund", funded partially by deposits from aquaculturists, to be used for remedying any environmental damage caused by aquaculture farms. In addition, the constitution of the Aquaculture Authority of India has been made for the regulation of shrimp aquaculture.

ENVIRONMENTAL ASPECTS OF AQUACULTURE

Environmental issues have become of increasing concern for several reasons. The first is recognition of the increasing pressure on resources in some coastal and inland areas. More attention has been given to the impact of aquaculture on the environment in recent years, partly induced by some well publicized "crashes" within the shrimp industry.

This has been accompanied by the publicity given to environmental and social issues surrounding aquaculture. This type of exposure has had a profound impact on the public's perception of aquaculture, changing it from the "blue revolution" which would improve the availability of cheap, affordable protein to poorer people, to that of being an environmentally unsound means to produce luxury food items for consumers in developed nations.

Australia is adopting integrated farming technologies in a watershed where water is an economic resource and is at a premium (i.e. use must be paid for). This situation is in contrast to that encountered in some countries in the region where traditional integrated farming practices are reducing their levels of integration, applying less species diversification and more emphasis on high-value products.

Balanced Resource Use

There has to be a balanced and more efficient use of resources. Obtaining conflicts. This situation can be particularly acute when the government's capacity to enforce laws and regulations is limited.

It appears likely that economic issues related to resource use and the "polluter pays" principle will be faced, in the next few years, by aquaculturists in some countries of the region.

Minimizing Negative Impacts

Guidance is emerging for the application of practices for improved environmental management, and strategies have been identified for the following:

* Technology and farming systems: In recognizing the importance of appropriate farming technology/system and management of inputs and outputs, special attention is given to the major resources used (i.e. feed, water, sediments and seed). At this level, management actions mainly involve farmers and input suppliers.
* Adoption of integrated coastal area management approaches: The importance of integrating aquaculture projects within existing ecological systems in coastal areas is increasingly recognized. This approach requires the consideration of proper site selection and the application of planning and management strategies that allow the allocation of resources among different users.
* Policy and institutional support: It is necessary to have a clear and supportive policy framework for aquaculture. Particular issues include aquaculture legislation, economic incentives and disincentives, actions for the public image of the activity, private-sector and

community participation in policy formulation, and increasing the effectiveness of research, extension and information exchange. Policy decisions and their implementation play a strong role in influencing the management possibilities at both the farm and local area levels.

The environmental assessment of an aquaculture area, rather than targeting individual projects, is now being used to identify opportunities for minimizing the impacts of the aquaculture sector within a particular area (e.g. in a bay, an estuary or a watershed). Such applicable measures are promoted through a wide range of instruments and should be brought together within the framework of an aquaculture development plan, ideally as part of an Integrated Coastal Management Plan. There is a growing awareness that environmentally friendly aquaculture makes good business sense, and this is particularly true for commercial producers. It is increasingly important when considering the perceptions of target export markets and provides an incentive for both the industry and supporting governments to further advance the adoption of environmentally and socially responsible farming practices through appropriate standards or codes of conduct. Recognized codes of practice require further development and implementation and are becoming particularly important for aquaculture products that are traded internationally.

AQUACULTURE AND ENVIRONMENTAL REHABILITATION

It is well known that inland aquaculture, particularly that within integrated agricultural farming systems, can provide many positive environmental benefits (e.g. water storage on small-scale farms, efficient use of farm resources). Perhaps less well known and publicized is the fact that coastal aquaculture, including shrimp culture, can also contribute to environmental improvement. For example, seaweed and mollusc culture can contribute to the removal of nutrients and organic materials from coastal waters. It can also provide an alternative source of income and employment for people involved in environmentally destructive fishing practices and other forms of habitat destruction. Other activities of note include the following:

- * Mixed aquaculture-mangrove systems are being used to restore mangrove habitats in some countries (e.g. Indonesia's tambaks or earthen ponds, Viet Nam's mixed farming systems);
- * Coral reef fish mariculture provides an effective alternative to destructive fishing practices in coral reef areas;
- * The rehabilitation of fish populations through stock enhancement; and
- * Aquaculture itself is also a technique for the effective monitoring of environmental conditions.

A recent review has suggested that aquaculture within saline inland areas can provide benefits through the use of previously degraded resources, contributing to environmental rehabilitation and poverty alleviation among communities whose livelihoods such damaged resources.

Aquaculture is also being integrated as a component of mangrove rehabilitation projects (as in the Mekong Delta) and within coral reef management activities (small-scale cage culture).

BIODIVERSITY, GENETICS AND AQUACULTURE

The potential impact of farmed aquaculture stocks, especially introduced exotic species, on wild fish and fisheries has been highlighted at many international meetings. Among the potential impacts identified are a loss of genetic resources within healthy or undisturbed populations and a dilution of the wild gene pool through interbreeding with escapees from aquaculture facilities.

Initiatives have been made recently to breed indigenous species for stock enhancement purposes as well as for aquaculture, but these are new initiatives and results that could guide development work are not available yet.

Aquafeeds and Feeding Strategies

When feeds are used in aquaculture, their contribution to production costs varies following the species and the system employed but can be as high as 50 percent of the total costs. Following concern as to the sustainable supply of fishmeal and fish oils, research has focused on identifying alternative, economically viable sources of proteins and lipids for incorporation into feeds. Improved management for more efficient feed use is another area of interest.

Some research (i.e. in Hong Kong) has taken a step further towards the development of environmentally friendly feed. Little information is available on the production of aquaculture feeds in Asia, but a recent estimate indicates a supply of some 1.9 million mt in Asia (excluding China), where 1.2 million mt were for fish and 0.7 million mt for shrimp. An estimate for China (1996) indicated a total aquaculture feed production of 5 million mt, but it was unclear if this figure also included farm-made feeds. The future prospect for fishmeal and oil for animal feeds is that global supply is expected to be stable (FAO, 1998).

A distinction must be drawn between the different feeds manufactured for aquaculture species, i.e. whether for high- or low-value species. The higher values of commodity species, or those reared for export, can justify the investment required for the manufacture, distribution, purchase and use of specially formulated feeds. However, much of Asian aquaculture involves smallholder-based production of low-value species, either for personal consumption or for local sale. In the latter case, the input cost of feeds is generally negligible and, if feeds are used, they usually consist of cheap, readily available materials.

INTERNATIONAL DEVELOPMENT AID TO AQUACULTURE

While Asia has received much of the international external assistance for aquaculture development, FAO (1998) has reported reduced levels in recent years. However, the recent trend for such support has been through assistance given to projects having a wider scope. These include as socially oriented projects, environmental rehabilitation projects, as well as integrated area development programmes, such as the Mekong River basin development programme, and the Bay of Bengal Programme (BOBP). In many of these projects, aquaculture is a component within a broad-scale development.

For the period 1988-1995, the Asian Region accounted for 65 percent of commitments made by funding agencies and donors to development assistance (which is different from actual disbursements) and 38 percent of the projects, while Africa was the recipient of 16 percent of commitments and about a quarter of the projects. The major beneficiaries during this period (i.e. countries receiving at least 1 percent of total aid) included India, China, Bangladesh and Mexico, which all received major loans from development banks, accounting for about 64 percent (about US$638 million) of external aid to aquaculture.

The two most common sources of credit for Asia's developing countries have been the World Bank (WB) and the Asian Development Bank (ADB). In 1997, WB provided US$409 million to South Asia's agriculture sector (a decrease of 26 percent compared to 1995), while the Asian Development Bank approved 632 loans, of which 10 percent went to the agriculture sector and 50 percent to the financial sector.

Other multilateral programmes that assist small farmers are the United Nations Development Programme (UNDP) (for agricultural development, including training and provision of experts) and the United Nations Development Fund for Women (UNIFEM) (provision of financial capital enabling women to establish small businesses, including aquaculture).

The United Nations Children's Fund (UNICEF), the United Nations Drug Control Programme (UNDCP) and others, including FAO, extend support to small farmer development programmes under various schemes (Technical Cooperation among Developing Countries (TCDC), Special Programme on Food Security (SPFS), Technical Cooperation Programme (TCP)). In addition, a number of donor agencies provide small-scale enterprises with development fund assistance. These include the United States Agency for International Development (USAID], the Australian Agency for International Development (AusAID), the Canadian International Development Agency (CIDA), the Danish International Development Agency (DANIDA), the Swedish International Development Agency (SIDA), the Norwegian Agency for Development Cooperation (NORAD), the Deutsche Gesellschaft für Technische Zusammenarbeit (GTZ), the Department for International Development (DFID, United Kingdom), the Groupe de Reflexion d'échange Technologique (GRET, France) and agencies from Belgium and the Netherlands. Assistance to research and development is also provided by the Australian Centre for International Agricultural Research (ACIAR), the Danish Cooperation for Environment and Development (DANCED), the Japanese International Cooperation Agency (JICA) and the other afore-mentioned agencies.

Assistance by Banks, Private Sector to Aquaculture Development

Private-sector investment for aquaculture development has tended to be limited to the culture of high-value species, where the systems are more capital intensive. The degree of personal involvement of the investor depends upon the scale and cost of the project. Small-scale projects may be financed directly by the owner or jointly with other investors, with the farm operating without recourse to external sources of finance. As the investment and operating capital requirements increase, it is often necessary for such entities to turn to external sources of funding (e.g. local or national banks).

For larger scale production units, particularly in cases where the investor has little or no direct control over day-to-day operations, such investment is often assessed alongside other potential opportunities. These circumstances can be quite volatile and depend on the relative performance of the investment(s). For example, the short-term nature of many of the shrimp farming investments has resulted in a lot of movement of investment capital, both into and out of the sector. The Asian Region does not possess a well-developed venture capital market that is willing to invest in new projects of potentially high financial risk. This situation severely limits the region's capability to develop and bring to the market new technologies without recourse to foreign investment.

This situation can also lead to a loss of opportunity, particularly where the fruits of research that have been achieved in the region benefit projects and investors elsewhere. Additionally, future prospects for private-sector financing are likely to be affected by the current weak economic situation in much of the region.

It is clear that many projects that were funded in the years before the economic crisis in Asia were not subject to the kind of scrutiny that would have identified high-risk investments. The banking reforms that are being undertaken in the region are likely reduce the level of financing available to aquaculture projects unless these can be demonstrated to be technically and economically viable. At the same time, the increasing costs of compliance with government regulations, as well as the trade agreements referred to, will increase the cost of entry into the commercial aquaculture sector.

NGO support to projects involving aquaculture has also increased, noting that there are thousands of NGOs in Asia. National development banks have tended to give a higher priority to projects producing high-value crops, although some, like India's NABARD, the Grameen Bank in Bangladesh, Bank Pertanian Malaysia, Bank Rakyat Indonesia, Bank of Agriculture and Agricultural Co-operatives of Thailand, and the Land Bank of the Philippines and others have developed loan programmes for small projects oriented towards rural development.

The private sector has played a major role in Asian aquaculture development, although principally for commercial rearing of higher value species. Due to the predominance of small-scale farms in some countries and the rate of development of intensive aquaculture, feed companies and other suppliers have proved to be a significant source of information for farmers.

Shrimp feed companies, in particular, frequently provide information and technical support as part of their customer services. Such information is usually aimed at increasing sales.

Sometimes, this can lead to the widespread recommendation of inappropriate practices, such as overstocking. On the other hand, it has been customary to organize seminars, with invited independent experts, to pass on new information to farmers. On balance, the impact of this type of information transfer has been positive for the production sector. Most feed companies also provide technical support services for both their current and prospective customers, which may include checking water quality, fish or shrimp health and providing advice on the farm. Some suppliers have introduced testing services for shrimp viruses.

The creation of farmers' groups has become more common in recent years, partly as a response to the pressures put on the production sector by external sources, but also with the realisation by the farmers that a strong professional lobby was needed to put aquaculture on the political agenda. This type of action has improved the exchange of views between the appropriate government bodies and the private sector and has also improved the awareness of wider issues among the members of such farmers' groups.

Within the region, much aquaculture investment has been, essentially, of a short-term nature. This position, when combined with the lack of local sources of venture capital, has meant that the region, with one or two exceptions, lags behind considerably in the private-sector development of new technologies. An exception to this was the establishment of a consortium of several private-sector companies for the development of a shrimp-breeding programme in Thailand. However, this might not have been established without the major involvement of a government agency that is a major shareholder in the venture. This position demonstrates the contrast with other shrimp breeding initiatives in the Americas that are mostly financed with private-sector funds, albeit using research results achieved through governmental support (universities, institutes and other public bodies).

CURRENT STATUS OF AQUACULTURE

Compared with fishing, aquaculture is currently of little commercial significance to the Pacific Islands, with one important exception, black pearl farming, which is virtually confined to eastern Polynesia. Elsewhere in the Pacific, considerable development is needed before aquaculture can be considered economically sustainable. Shrimp (Penaeus spp.) farming has been a focus of commercial development in several islands with varying degrees of success; tilapia (Oreochromis niloticus) aquaculture has entered the subsistence economy in some areas, and seaweed (Kappaphycus spp.) is a future commercial export prospect. The culture of other marine and freshwater species is, however, generally still at the experimental or "backyard" stage.

The expansion of aquaculture in the Pacific will depend on providing better production methods for species currently being farmed, and techniques for propagating and growing the "new" species described above. These methods and techniques should be simple and flexible so that they can be adapted to the context of the Pacific Islands environment and to the market constraints (local and export markets). This approach should favour systems integrating fisheries and mariculture with low investment and operating costs and simple technical production processes. This should be done in association with pilot commercial-scale operations to test and demonstrate the economic viability of the methods proposed. This will require research combined with assistance, training and education programmes.

Pacific Island nations have many attributes that favour development of aquaculture and stock enhancement in the coastal zone. These are as follows: a great diversity of coral reef species which are in high demand, proximity to major aquaculture and seafood markets in Asia, availability of suitable growout sites in pristine habitats, geographic conditions which favour restocking and stock enhancement, a relatively inexpensive labour force, and a tradition of working with marine resources.

Although these confer many advantages on the region in terms of aquaculture development and stock enhancement, there are still several constraints for such enterprises in the Pacific, which include limited domestic markets, high added-value export markets targeted, transport problems, socio-economic factors, fragile habitats, limited fresh water, and cyclones. Some of the best opportunities for aquaculture development in the Pacific are in the aquarium trade (coral reef fish, hard corals, soft corals), the live seafood markets (e.g. groupers, spiny lobsters, abalone, crabs) and the pharmaceutical industry (e.g. algae, sponges, soft corals). In all cases, the products are of high value and can be grown in small areas with relatively simple technology.

This review covers the insular Pacific as defined by the work-area of the Secretariat of the Pacific Community (SPC).

Compared to fishing, aquaculture is currently of little commercial significance in the insular Pacific, with one important exception-black pearl farming-and this is virtually confined to eastern Polynesia. Elsewhere in the Pacific, considerable development is needed before aquaculture can be considered economically sustainable. Shrimp (Penaeus spp.) farming has been a focus of commercial development in several islands over the past 30 years, with varying degrees of success; tilapia (Oreochromis niloticus) aquaculture has entered the subsistence economy in some areas, and seaweed (Kappaphyceus spp.) is considered a future commercial export prospect by the region. The culture of other marine and freshwater species is generally, however, still at the experimental or "backyard" stage.

Aquaculture is a relatively new development in the region, and in most Pacific Islands where it has been attempted; its history goes back less than 30 years. There is no fund of traditional knowledge for culturing fish and shellfish, just catching them, except in very specialized instances and areas. There is thus no great resource of aquacultural skill or infrastructure. This steep development path has perhaps not been taken into account in some development projects, which have often had unrealistic short-term aims and lacked follow-up.

Despite the comparatively minor penetration of aquaculture into Pacific Island economies, and despite the loss of interest by most of the international development community after many short-term project failures, several Pacific Island governments have accepted the challenge. They recognize that expansion in capture fisheries is limited, and have made substantial investments in freshwater aquaculture and mariculture, often in concert with external sources of development assistance.

Pacific Islanders cannot turn away from the sea. It is the greatest resource they have.

The Potential Importance of Aquaculture to Pacific Islands: A major problem facing most of the island nations in the Pacific is that they have relatively few opportunities to generate income (Adams, 1998). The economies of most Pacific countries are limited due to small landmasses, few terrestrial resources and low numbers of inhabitants. To ensure further development, island nations must make the most of the one important resource they all have-the sea. Through the joint efforts of the Forum Fisheries Agency (FFA) and the SPC Oceanic Fisheries Programme, island nations are deriving major inputs to their economies by fishing for tuna, or by selling access rights to tuna, within the large maritime zones under their control. However, valuable, sustainable harvests are also possible from inshore waters and coral reefs.

The potential for increased well-being of coastal communities from the responsibleuse of their inshore marine resources arises because the inshore habitats surrounding Pacific nations support a great diversity of economically important species (Wright and Hill, 1993; Dalzell et al., 1996; Bell and Gervis, 1999). Traditionally, these animals were harvested at subsistence levels. More recently, however, development of export markets has provided coastal communities with opportunities to earn income from the inshore fisheries species. Unfortunately, the transition from a subsistence to a market economy has usually been far from ideal: chronic overfishing has occurred in some areas. In such places, there are now too few of the prized animals to sustain reasonable harvests. Destructive fishing methods have compounded the problem by degrading some habitats to the point where they cannot support the valuable species (McManus, 1997). Pacific Island countries now recognize that aquaculture provides one of the few long-term, sustainable, ways of deriving benefits from inshore fisheries resources (Williams, 1996). This view of aquaculture as a priority area for continued sustainable development was reinforced as part of a consensus member country statement arising from the 2nd SPC Fisheries Management Workshop in 1998:

> *"...Regional fisheries managers have focussed on establishing regimes to sustain inshore fisheries. This is supported by a strategy to divert demand and fishing pressure to alternative activities, mostly to offshore fishing and into aquaculture.*

Development of fisheries management varies from country to country reflecting the differing stages of economic development and levels of need. In some countries the need is to encourage economic activities and to generate income for rural villages; in other countries the need is to restrict or limit fishing. Yet in other countries the need to involve all stakeholders in the management system has evolved into community-based decision-making and control. There is now broad acceptance that marine resources cannot be managed in isolation from other users, or by one government agency so that an integrated and co-ordinated approach should be taken. In many circumstances, because of the smallness of the islands, an island system-management approach is the desirable option.

Aquaculture, as an alternative activity, is still at a preliminary stage of economic development in most PIC3, but is of enormous future significance. For aquaculture to realise its full potential to the economies of PIC in a sustainable way will require a considerable degree of international support. PIC have endorsed a strategy to harness and prioritise such support at the regional institutional level. Several PIC already devote significant national resources to this subsector and this trend will continue as benefits are realised..." Advantages of Pacific Islands for Aquaculture: Pacific Island nations have many attributes that favour development of aquaculture and stock enhancement in the coastal zone. These include:

- o A great diversity of species associated with coral reefs that are in high demand for:
- o the aquaculture and seafood markets in Asia (e.g. napoleon wrasse, groupers, sea cucumbers, spiny lobsters, trochus, pearl oysters, giant clams, green snail);
- o the marine aquarium trade (e.g. clownfish, angelfish, hard corals, soft corals, giant clams); and
- o the pharmaceutical trade (e.g. algae, sponges, soft corals, sea horses).
- o Proximity to the major aquaculture and seafood markets of Asia-flight times are short enough to ensure that many species can be shipped alive to Asia.
- o Availability of suitable growout sites in pristine habitats-coral reef lagoons create the calm conditions essential for culture of many species. The favourable environmental conditions should be the opportunity to develop green label products to get better prices on the international market.
- o Geography that favours restocking and stock enhancement-most Pacific countries are small islands, or groups of small islands, surrounded by deep water. Cultured juveniles released into the inshore waters of island ecosystems cannot emigrate, and are therefore relatively easy to recapture.
- o A relatively inexpensive labour force-expectations for financial return on labour are low in many Pacific countries relative to developed countries.
- o A tradition of working with marine resources-coastal communities are already familiar with the basic biology of many species.

Constraints to Aquaculture in Pacific Islands

Although the attributes listed above confer many advantages on the region for development of aquaculture and stock enhancement, there are also several constraints to such enterprises in the Pacific. Many of these have been identified previously by Uwate and Kunatuba (1983), Munro (1993) and Bell and Gervis (1999). They include:

- o Limited domestic markets. Local markets for the fresh products of aquaculture in the Pacific are small, and with the exception of very limited opportunities in the restaurant trade, usually offer low prices. Thus any large-scale aquaculture development in the Pacific catering to the trade in seafood will depend heavily on export markets.
- o High added-value export markets targeted. These are most often regional and fluctuating markets (e.g. live reef fish). Thus any aquaculture development will depend on the capacity to follow the market trends, and to fulfil the demand in time.
- o Transport problems. The high cost of shipping in the Pacific adds considerably to the cost of producing and exporting aquaculture products. Poor internal transport services restrict opportunities to grow perishable products in remote locations, and limited international air connections inhibit continuity of supply to export markets. Transport arrangements dictate that species cultured for export need to be of high value and low weight. Alternatively, the products must be non-perishable, e.g. bêche-de-mer (processed sea cucumbers) or frozen, so that they can be shipped by sea.
- o Social and economic factors. Many of the smaller island nations lack the infrastructure, capital and skilled labour required to implement aquaculture, particularly where hatcheries are involved. Sustained assistance from developed countries is needed to implement and operate stock enhancement programmes until they become self-funding (Bell, 1999a). The traditional

marine tenure systems in place in many countries (Ruddle et al., 1992) also add complexity to the process of negotiating access and tenure to sites for aquaculture.

- o Fragile habitats. Coral reef ecosystems on many of the smaller island nations have evolved in a nutrient-poor environment. Additions of nutrients, e.g.through uneaten and undigested formulated diets for carnivorous fish in cage culture (Beveridge, 1987; Landesman, 1995; Stewart, 1997), can be expected to change the ecosystem in favour of algae and herbivores. Such changes are likely to be unacceptable, particularly to the tourist industry. This constraint is particularly relevant to lagoonal habitats, but would not apply to locations that have good flushing to the open ocean.
- o Freshwater is limited, except for the large islands of Melanesia, which have extensive river systems. Prospects for freshwater aquaculture are thus limited. Even in areas with significant fluvial development, the indigenous freshwater ichthyofauna is generally unfavourable for economic culture and species for freshwater aquaculture have been imported.
- o Cyclones. Countries in the cyclone belt can expect aquaculture installations to be damaged intermittently by large swells and strong winds.

Aquaculture Systems: Profitable aquaculture of penaeid shrimps and blacklip pearl oysters has now been established in some areas of the Pacific by commercial interests. Stand-alone enterprises producing penaeid shrimps for export markets are firmly established in New Caledonia and Fiji and were so in Solomon Islands until recently. These enterprises are applying technology developed originally in Japan, Taiwan Province of China and France, and now common place throughout the tropics.

A large, sustainable, industry for culturing pearls using the blacklip pearl oyster (Pinctada margaritifera) has been established in the Tuamotu Archipelago, French Polynesia, and on a couple of atolls in the Cook Islands (Fassler, 1995).

In French Polynesia, the value of cultured pearls exceeds US$150 million per year. In Cook Islands, the industry is currently worth US$5 million and is the second largest source of revenue for the country after tourism.

Black pearl farming in French Polynesia and Cook Islands, and shrimp aquaculture in New Caledonia represented more than 98 percent of the total value of aquaculture production estimated in 1996.

Blacklip Pearl Oysters: Small-scale culture of pearl oysters is under way in Fiji, Marshall Islands, the Federated States of Micronesia, Solomon Islands, Kiribati and Tonga. In some places, e.g. Kiribati, development is based on spat produced in hatcheries, whereas in others, e.g. Solomon Islands, development is geared towards finding ways that coastal villagers can catch and grow wild spat (Friedman et al., 1998). Current research is concentrating on assessing the economic viability of pearl farming in Solomon Islands, Fiji and Kiribati, and comparing growth, survival and pearl quality of oysters derived from wild and hatchery-reared spat.

Giant Clams: Small-scale enterprises in Solomon Islands, Palau, Marshall Islands, Cook Islands, Tonga and American Samoa supply five species of giant clams (Tridacna crocea, T. derasa, T. gigas, T. maxima and T. squamosa) to the marine aquarium trade (Foyle et al., 1997; Hart et al., 1998). Production of giant clams for enhancement of wild stocks is also under way in Solomon Islands, Fiji, Cook Islands and Western Samoa (Bell et al., 1997a; Bell, 1999b). Several of these countries also have the capacity to produce giant clams for the sashimi market in Okinawa, and as a live product for markets in China, Hong Kong Special Administrative Region and Taiwan Province of China (Bell et al., 1997b).

Sea Cucumbers: Research has commenced to assess the viability of producing sea cucumbers in hatcheries for enhancement of wild stocks. There are three steps in this process: developing methods for cost-effective mass production of juveniles, learning to release the cultured juveniles in ways that maximize their survival, and evaluating the economic impact of releasing cultured juveniles into existing fisheries. Currently, the focus is on development of methods for the mass rearing of Holothuria scabra, H. fuscogilva and Actinopyga mauritiana, three of the most valuable sea cucumbers in the region. To date, ICLARM has demonstrated that H. scabra is relatively easy and cheap to rear (Battaglene and Bell, 1999), and that A. mauritiana grows relatively rapidly at high densities (Ramofafia et al., 1997). Initial research on H. scabra indicates that this species has much potential for stock enhancement.

Other Species: Technology for propagating and releasing cultured juveniles of green snail and trochus has been transferred to the Pacific through projects in Tonga (JICA) and Vanuatu (Austrailian Centre for International Agricultural Research, ACIAR). The Overseas Fishery Cooperation Foundation (OFCF) is also implementing a stock enhancement programme for both species in Solomon Islands. Production of the marine alga Kappaphycus is well established by coastal villagers in Kiribati and Fiji, and sponges are being cultured in the Federated States of Micronesia. Milkfish are being cultured as live bait for the tuna industry in Guam, and there is considerable interest in this activity by several other countries.

Policy and Institutional Framework: The independent Pacific Islands (Cook Islands, Federated States of Micronesia, Fiji, Kiribati, Marshall Islands, Nauru, Niue, Palau, Papua New Guinea, Samoa, Solomon Islands, Tonga, Tuvalu, Vanuatu) generally lack any specific provision for aquaculture in the legislation, but many include a statement about aquaculture in national development plans.

Whilst prospects for inland aquaculture are limited by geography, the custom of communal tenure of coastal marine areas may be incongruous with private-sector farm ownership in many areas, unless the development is carefully managed. Tonga is currently developing a comprehensive legislative basis for future aquaculture development, whilst the Cook Islands has adapted traditional systems into a legal basis for pearl farm management. Other countries are taking a more ad hoc approach and trying to adapt land-based systems of state leases for mariculture development, or encouraging development only by traditional reef custodians.

Fiji: Aquaculture was first started in 1953, when tilapia were introduced as a protein source for pig farming. The first directed efforts occurred after the United States Peace Corps and a JICA project assisted the Fiji government in developing freshwater aquaculture methods in the 1980s. Varied success was achieved with shrimp, Kappaphycus, oyster, mussel, Macrobrachium, carp and tilapia, but now major investment is promoted by the government through a Commodity Development Fund. Currently, there are three shrimp farms and hatchery; 20 milkfish ponds for longline bait; and one industrial, seven commercial and 215 subsistence farms and six hatcheries that produced 243 mt of tilapia in 1998. Eucheuma cottonii is cultured by a total of 182 farms, producing an estimated 1 500 mt in 1999. Experimental pearl farming is conducted, and there is one commercial farm that has been operating for two decades. There is also a giant clam hatchery, the Naduruloulou Government Freshwater Aquaculture Research Station, the Makogai Government Mariculture Research Station, and the University of the South Pacific (USP) Aquaculture Teaching-School Teaching Ponds (E. Ledua, 1999, unpublished data).

Guam: There was an increasing trend in annual aquaculture production between 1990 and 1995 (Clarke, 1997). In 1995, aquaculture production was 205 mt valued at US$1.6, derived mainly from the farming of tilapia, marine shrimps, Chinese catfish and milkfish. The potential for the development of other species (giant clams, top shell, striped mullet and groupers) is being explored by Guam-based research facilities.

Kiribati: Eighty hectares of mikfish ponds, originally set up to provide livebait for tuna pole-and-lining, have been operating for several decades. Kappaphycus alvarezii has been cultured in the Phoenix, Line and Gilbert groups for 15 years. There is an OFCF project on sea cucumber rearing and an ACIAR project on pearl farming (T. Tekinaiti, 1999, unpublished data).

Nauru: Milkfish been farmed for at least one century, but competition from introduced tilapia (O. mossambicus) occurred in Buada Lagoon and culture lapsed. It has recently been revived, with O. niloticus and milkfish being raised in 11 ponds (F. Alefaio, 1999, unpublished data).

Northern Marianas: Aquaculture is limited to raising penaeid shrimp and tilapia (Clarke, 1997). Annual production is estimated at 1 250 kg for marine shrimp, 200 kg for freshwater prawns and 4 200 kg for tilapia. There are fewer than three commercial farms. The total value of products was estimated as US$25 000 in 1996.

Nouvelle Caledonie: Shrimp production achieved 1 500 mt for the first time in 1998 (Labrosse et al., in press), after disease problems which occurred in the early 1990s. The value of marine shrimp production is about US$10 million. About 45 mt of oysters (Crassostrea gigas) are produced for the local market. Experimental culture of local oyster and giant clam is also in progress (Etaix-Bonnin, 1999).

Palau: The Micronesian Mariculture Demonstration Centre pioneered giant clam culture, as well as trochus and soft corals.

Papua New Guinea: Aquaculture started 40 years ago, with several aquaculture stations along the coast and highlands to encourage subsistence culture, mainly of Cyprinus carpio. There are 300 carp farms in operation. Trout were introduced in the 1940s, and the Kotuni Trout Farm was in operation from 1973-1984. There are three newer farms, but only two are currently operating, with a production of 15 mt. A hatchery was started in 1996.

Culture of barramundi (Lates calcarifer) has been started in Madang on the site of a failed 9 ha pond originally used for redclaw (Cherax quadricarinatus), but production has not yet been established (J. Wani, 1999, unpublished data).

Samoa: Farming trials were conducted on tilapia in 1954 by the SPC and on Macrobrachium in 1971 by the FAO, but were not successful. Seaweed, giant clam, green mussel and redclaw farming were also tried. A new national economic strategy promotes aquaculture, and this is being actively developed by the current AusAID village fisheries extension project using tilapia, mullet and giant clam.

Solomon Islands: There are two shrimp farms and several village-based enterprises rearing giant clams and hard corals for the aquarium trade and a demonstration black pearl farm in the Western Province. The ICLARM Coastal Aquaculture Centre has also developed methods for the propagation of sandfish (Holothuria scabra) and the capture and culture of wild postlarval coral reef fish for the aquarium trade. Between 1996 and 1999 there was an OFCF project on green snail & trochus (E. Oreihaka, 1999, unpublished data), but the future of the aquaculture-oriented Institute of Marine Resources of the University of the South Pacific is uncertain.

PROSPECTS FOR FURTHER DEVELOPMENT

Some of the best opportunities for development of aquaculture in the Pacific are in the aquarium trade and live seafood markets (e.g. napoleon wrasse, groupers, sea cucumbers, spiny lobsters, trochus, pearl oysters, giant clams, green snail, abalone, crabs, clownfish, angelfish, hard corals, soft corals) and the pharmaceutical industry (e.g. algae, sponges, soft corals, sea horses) (Bell and Gervis, 1999; Bell, 1999c). In all cases, the products are of high value and can be grown in small areas with relatively simple technology.

Initiatives by FAO, the World Fish Center (ICLARM), the Center for Tropical and Subtropical Aquaculture (CTSA) and bilateral donors have concentrated on establishing the culture of pearl oysters outside eastern Polynesia, developing small-scale aquaculture enterprises for other species, and providing basic training in aquaculture and stock enhancement to fisheries staff in several of the countries.

The proceedings of the workshop entitled Present and Future of Aquaculture in the Pacific organized jointly by the South Pacific Aquaculture Development Project (SPADP) and the Japanese International Cooperation Agency (JICA) in Tonga in November, 1995 (Anon., 1996) presented the status of aquaculture in the region.

The continued expansion of aquaculture in the Pacific will depend on providing better methods of production for species currently under cultivation, and techniques for propagating and growing the "new" species described above (Bell, 1999c). These methods and techniques should be simple and flexible enough to be easily adapted to the context of the Pacific Islands environment and to the constraints of local and export markets. This approach should promote systems integrating fisheries and mariculture, with low cost methods of production). This should be associated with pilot commercial scale operations to test and demonstrate the economic viability of the methods proposed. This will need research coupled with assistance, training and education programmes.

In recognition of these needs, three organizations (SPC, ICLARM and USP) have recently joined forces to produce a "Regional Strategy for the Development of Aquaculture". Under this strategy, SPC will be the regional focal point for aquaculture and willconvene regular meetings of island nations to identify needs, determine priorities and put organisations and individuals in touch with each other. ICLARM will undertake long-term research to devise and test economically and environmentally sustainable methods for restocking, stock enhancement and farming; and USP will developdegree and diploma course components and aquaculture vocational training, and contribute to research through higher degree programmes. The other functions necessary for the expansion of sustainable aquaculture in the region, e.g. marketing, legislation, environmental protection and quarantine, will be integrated progressively through association with the Strategy.

This Pacific Islands Regional Aquaculture Strategy represents an opportunity to reinforce inter-regional cooperation based on research, training and information exchanges (including cooperation through NACA with Southeast Asian countries). It may promote more investment from Asia and better conditions for access to Asian markets.

FISH, THEIR "FRIENDS," AND THE FREE MARKET

Although marine fish, as residents of interconnected bodies of water that span the globe, are continually mobile and nearly impossible to localize or claim, as a resource for human consumption they have become a source of political debate and conflict between very localized states and groups of people. Currently, as the threatened state of global fisheries resources becomes increasingly prominent, the lines for disagreement, conflict, and negotiation are diverse and run the spectrum from international organizations to groups of different economic and social strata at the level of coastal communities. To people around the world of varying wealth, nationality, economic system, culture, and position of power, fish play a crucial role in their tastes, traditions, and needs for survival. This, more than what needs to be done ecologically to restore the world's fish resources, is what makes the situation so complex.

One effort to simultaneously resolve the seemingly disparate elements of the problem-the economic, the environmental, and the developmental-is the concept of sustainable development. In the context of fisheries, this concept entails being able to manage the pressures for the production and trade of fish, the strain this places on the resource base, and the needs of those people that depend on the resource for livelihood and food security to develop in an equitable manner. As ebbs or flows in one sphere impact the security of the others, solutions in no one sphere can be seen as representing a sustainable solution if the other two are neglected.

Of so much importance to so many people around the world, individuals, organizations, and states are searching for ways to alleviate the intense pressures on fisheries resources without completely sacrificing the interests of humans in using the resource for consumption and a source of livelihood. One current example of this is the proposal among states and international conservation and sustainable development organizations to use subsidy reform in the World Trade Organization (WTO) to pursue sustainable development objectives. While subsidies are a crucial contributor to the overfishing that occurs in many of the world's most important fisheries, the questions is whether or not such a reform can deliver on all three pillars of sustainable development.

Subsidy elimination or reform in the fisheries sector through the WTO holds several legitimate possibilities for better resource use and alleviation of perverse economic incentives in fisheries products production that suggest advancement in important aspects of human development in developing coastal countries. This paper intends to explain, though, the limitations of this action and the possible threats to development and resource use in developing countries posed by such an endorsement of free market principles, especially given certain intranational realities of many of the most fish-wealthy developing countries.

CURRENT STATE OF DEPLETION IN FISHERIES RESOURCES

The first step towards an accurate conception of current global fish-use, and possibly the least debatable aspect of the issue, is recognition of the precarious biological and environmental situation of the world's fish stocks in the face of increased global consumption of fish in the past 30 years. Since 1973, global consumption of fish has doubled. Growth in demand for fish as a source of protein in the developing world has accounted for ninety percent of this growth, while developed countries' share in world fish consumption has decreased. Fish is consumed most highly by far in Asia, and next highly in Europe.

Fish consumption is generally higher in developed countries, but may play a larger role in food security in developing countries. At the same time, populations of fish in almost every part of the globe have declined to levels well below those they have historically and naturally held. The World Wide Fund for Nature (WWF) reports that 73 to 75 percent of the world's major fisheries are overexploited, fully exploited, or recovering from depletion. An article from the United Nations Food and Agriculture Organization (FAO) reports that, "Judging from known fish stocks and resources of traditional fisheries, the total marine catches from most of the main fishing areas in the Atlantic Ocean and some in the Pacific Ocean would appear to have reached their maximum potential some years ago." Global demand remains high, and the world's fisheries seem unable to support it in any sustainable way.

The link between increased global demand for fisheries products and the current level of stock exploitation is the dramatic increase in production that has occurred on an international level in the same time period. While 44 million tons of fish were captured for food in 1973, 65 were in 1997. Since the 1980s, developing countries have taken the lead in wild fish capture and now account for over 70 percent of global fish production. After such high levels in the last 30 years, though, wild fish production has stagnated; the state of exploitation in main fishing areas has actually remained more or less unchanged since the early 1990s. The world's fish stocks cannot continue to sustain such intensive levels of capture, a reality which has serious implications for both those on the production and consumption side of fishing.

The international trade of fish is also an important dynamic in the current situation. About 40 percent of world output by value of fish was traded across international borders in 1998. Since the early 1970s, the direction of fish trade has experienced an "about-face:" developed countries were net exporters of 818,000 tons of food fish in 1973 but, in 1997, were net importers of over four million tons of food fish.

More than half of fishery exports originate in developing countries, while developed countries account for 84 percent of world fishery product imports. Fish products are vital exports for many diverse economies, but the top exporters of fish and fish products are Thailand, the United States, Norway, China, Denmark, and Taiwan (in that order). Imports into the three largest importing blocs, Japan, the EU, and the United States, accounted for over three-fourths of the world market in international fish trade in 1994. As will be addressed, although fish trade is important for many, it is hardly on equal levels or terms, and benefits are rarely bestowed on those to whom the resource matters most for food and livelihood security.

FISHERIES: AN OPEN ACCESS RESOURCE

Due in large part to the way it has traditionally been managed, marine fish has usually been seen and used as an open access resource, an in many ways endemic characteristic which has been the base of its dynamics of use, management, and exploitation. The term refers to a situation where no single user has to pay for access to a resource, nor does that user have exclusive rights to the resource or the right to prevent others from using it. Physical characteristics of the oceans, and the fish they contain, such as scale, interconnectedness, and mobility of fish stocks, contribute to this open access nature. As such, they have usually defied national borders and been logistically difficult or impossible to monitor and govern. Theoretically, this can lead to a "tragedy of the commons," where "the open-access nature of the fisheries allows more and more people to enter the fishing sector, which leads to the economic (and possibly biological) over-exploitation of the resources."

Even after the resource begins to decline, producers will continue to increase their efforts until operating costs can no longer be covered. By this time, the stock of the resource may be seriously depleted. The common-access nature, which has long been a characteristic of fisheries, is the foundation on which many of the causes of current problems in the sector are based.

Root Global Causes of Fisheries Resource Depletion

The simultaneous, and highly linked, realities that too many fish are being caught and the global demand for fish is unsustainably high are largely seen to be the result of four root causes: population growth, growth in incomes, urbanization, and advancements in fishing technology.

Population growth has contributed to an increase in demand for fish by increasing the number of people who demand fish. Income growth has played a role because, as incomes increase, wealthier people's tastes change from preferring low-value to high-value fish; more and more low-value fish are then diverted "from poor people's plates into the stomachs of expensive carnivorous fish, which will in turn end up on the plates of wealthier people." A 2003 report by the World Fish Center found that population growth and income growth were "widespread structural phenomenon," especially in the developing world, which drove increased demand and consumption in the last three decades.

Urbanization has also impacted the structure of fish demand because, as the "rapidly expanding" group of poor people living in larger cities throughout the world increases, low-value fish is their only affordable source of animal protein. On the production side, developments in technology that have allowed fishermen to catch more per unit input and seek out fish in deeper waters have kept production costs down and allowed production levels to increase.

Roman Grynberg and Martin Tsamenyi, for instance, cited constantly improving technology as a fundamental cause of fish stock depletion. The interaction of these structural forces of increasing incomes, populations, and urbanization, as well as advancements in technology, have lead to the over-use of the resource, which is, as is currently being witnessed, limited in its capacity to respond to and accommodate them.

MANAGEMENT AND SUBSIDIES

The very reality that the root causes of the over-use of fish stocks are structural makes it nearly impossible to completely alleviate the pressures of rising demand, increasing abilities to pay for high-value species, and more effective ways to capture fish on an open-access resource. But people, organizations, and governments have responded to these realities in many ways, their responses sometimes improving the situation, but often exacerbating it. Two of the most prominent and important responses have been subsidies and management schemes. The first, subsidies, occur when trade measures directly influence the income or reduce the costs of production for industry. Management schemes, the second, encompass the ways in which fisheries are governed and monitored. In practice, the way this is done often highlights a lack of effective management. Subsidies and management, which can be seen as responses by individuals, groups, or governments to the root causes of fisheries depletion, are also the areas where the most opportunities for change in the way fisheries resources are used exist.

Effective management regimes are considered by many to be key in the sustainable use of fisheries resources. Theoretically, effective management encourages a sustainable use of resources through providing conservation regulations and incentives for responsible use of the resource. They are especially beneficial under the pressures of international integration; responses by producers to the economic incentives from trade liberalization, for example, can be better regulated when effective management schemes exist so that over-exploitation and destruction are less likely to occur. Unfortunately, the prevalence of such management has been scant in real-world fisheries. As the United Nations Environment Programme (UNEP) reports, "very few fisheries management systems have demonstrated the ability to keep catches below levels that put pressure on stocks."

As noted in a study by the World Conservation Union (IUCN), "national and international progress on trade expansion and liberalization outpaces progress on fisheries management and articulation of sustainable development strategies." Fisheries management, crucial to sustainable use of the world's fisheries, has been difficult to implement effectively and has often left fisheries open to influences from other sources such as state trade measures.

Subsidies, one of these trade measures, have played a proven role in aggravating the causes of fisheries depletion. Theoretically, subsidies can lead to overuse of a resource by altering the behaviour of producers. Revenue-enhancing or cost-reducing subsidies, such as those that will be examined in this paper, increase marginal profits at each level of effort and therefore theoretically lead to an increased overall effort. This may have the short-term effect of creating additional economic rent for producers, but either new entrants or increased effort by existing producers, also encouraged by the subsidy, will shift the level of effort so that rent is eventually dissipated. Assuming that the management program does not effectively impose a sustainable level of resource use, cost-reducing and revenue-enhancing subsidies will push the level of overcapacity and overall effort even further than would an open-access resource in the absence of subsidies. The main theoretical effects of cost-reducing, rent-increasing subsidies are increased capacity, a delay in exit by producers from the sector, and increased effort by producers. These effects can cause the overuse of a resource, which can have many other significant impacts such as threats to food, livelihood, and resource security.

FISHING SUBSIDIES

Types Used by National Governments

In the real world of governments, fishers, fisheries resources, and the incentives that link them all together, a subsidy can take many forms. Fishing subsidies currently account for large portions of many industrialized country budgets; a 1998 World Bank study estimated annual budgeted expenditures on global fisheries subsidies to be between $14 and $20 billion, and that subsidies account for 20 to 25 percent of the sector's revenues. One type of subsidy, domestic assistance programs, form a clear example of government endorsement of the fishing sector. These programs are defined as budgeted assistance to fishing production, and often enhance operations and capacity. Japan, one of the biggest subsidizers of fishing activities in the world, has used domestic assistance programs to recruit young fisherman, give aid to fish cooperatives and boat owners, support marketing, and encourage price stabilization.

A 1998 World Bank report estimated that Japan budgets $270 million a year for these types of programs, and a total of $750 when aid for fishing vessel insurance is included.

The European Union (EU), another major subsidizer, supports such structural programs as fleet renewal and modernization, processing and marketing, aquaculture, maintenance of port facilities, and genetic product promotion in their fisheries sector. In addition, it pays for market supports to the sector in the form of a minimum import price program and measures to support price floors. Costs of these programs to the EU in 1996 were estimated to be about $530 million. While estimates of exactly what types of domestic subsidy programs the Chinese government uses are hard to determine due to scant data, they have clearly aggressively promoted the expansion of their fishing sector in the last 15 years (and have become the world's leading producer) and are estimated to budget between $500 and $750 million annually for domestic subsidies. While other countries subsidize their domestic fishing industries, they have comparatively small budgets for such programs as so do so on a smaller scale. The shear magnitude of the domestic subsidies in industrialized countries suggests the possible impacts on the behaviour of fishers their presence may cause.

A less obvious, but yet potentially just as influential, type of subsidy occurs when a country pays the access fees for their fleets to fish in foreign waters. Unlike domestic subsidies, these payments are not made directly to fishers or even domestic programs, but to the foreign governments who have jurisdiction over the fish stocks. The EU is one of the major subsidizers of distant-water fishing through its numerous access agreements with developing countries, mostly in West Africa. For example, the EU signed a five-year access agreement with Mauritania in 1996 that lifted an EU embargo on fishery imports from Mauritania and specified EU payments of almost $350 million over five years to the Mauritanian government. In return, the Mauritanian government granted increased access for EU fishing vessels in their waters, authorization of higher EU total harvests, and the specific allowance of EU-directed fishing of highly valued squid and octopus. The monetary benefits to the "cash-strapped" nation were described as a "windfall;" in consequence, the number of eligible EU boats rose from 165 to 240 and allowable EU harvests from 76,050 to 183,392 tons.

The EU has similar agreements with other West African nations, as well as others in East Africa, the Indian Ocean, and South America. Another distant-water fleet subsidizer, Japan, currently budgets about $200 million a year on distant water arrangements with foreign countries and foreign fisheries assistance programs (which often help secure fishing rights). These payments went mostly to developing countries in the Pacific Ocean and elsewhere. The subsidization of distant water fleets from industrialized nations can have serious effects on the coastal communities of the developing countries with whom these agreements are typically made.

Possible Impacts of Perverse Subsidization

The impacts of government subsidies of fishing are at least as complex and numerous as the types of subsidies themselves, but in the context of sustainable development those related to market access due to trade measures and resource, livelihood, and food security are the most relevant. The trade implications of the subsidization of the fishing industry arise when the lower costs producers see as a result of the subsidies repress world prices for fish products. Those producers, who are often from developing countries that cannot afford subsidies, are then excluded from the market because they cannot afford to produce at the world price.

For example, when Japan subsidizes its fleets to fish in the waters of Chile, Japanese fishers bring the fish back to Japan and sell them on their domestic market. Chilean fishers try to enter the Japanese market and sell these same Chilean fish to Japanese consumers, but are kept out of the market because their costs are higher and the prices on the Japanese market are too low.

In effect, Japanese fishers push Chilean producers out of the market for the sale of their own fish. This example illustrates the way in which subsidies in the fishing sector can exclude those who are outside the national subsidy regime from participating in the market.

Due to numerous examples where the presence of subsidies seemed to increase the capacity in the sector and, in turn, deplete available resources, fishing subsidies have also been blamed for a tendency to encourage overcapacity and overfishing. The case of stock depletion in Canada's Northwest Atlantic fisheries in the 1950s and 1960s illustrates how subsidies can increase capacity and lead to depletion. Between 1954 and 1968, Canadian subsidies increased the capacity of the Northwest Atlantic offshore fishing fleet by more than 18 times, creating twice as much capacity as could capture fish if resources were to be used sustainably.

In 1970, the Canadian government "acknowledged explicitly that its subsidies for vessel construction and modernization over the previous two decades had led to the rapid expansion of larger vessels, creating serious overcapitalisation." An Organization for Economic Cooperation and Development (OCED) study reported that subsidies also contributed to overcapacity in the fishing sector in many member states: New Zealand in the 1970s and early 1980s, Spain's Galician fisheries in the 1980s, Norway in the 1960s, the EU during the 1980s, and the United States (US) in the late 1970s and 1980s. These examples of the history of fishing fleet subsidization in industrialized countries and the resulting depletion of their stocks illustrates how overcapacity in a sector can encourage overuse of a fragile resource and lead to its exhaustion, posing threats to the surrounding ecosystem and human environment.

In addition to threatening the resources on which humans depend for their incomes, fishing subsidies may pose direct threats to crucial aspects of human and national development such as livelihood and food security. These threats are seen most strikingly in the interaction between developing nations and the industrialized countries with whom they sign access agreements. Frequent realities of EU-West African access agreements, for example, include a disregard of African fishing regulations and international agreements by distant water fleets, an inability by African states to monitor activities or catch levels of EU fleets, a lack of direct linkage between access payments and money spent on fisheries-related activities, a lack of enforcement of management efforts by African governments due to a fear of losing compensations, and the challenge of depletion of valuable fisheries stocks. There are also often many adverse effects on African coastal communities such as threats to the livelihoods of artisanal fishers and food security of those who depend of local fish stocks for protein intake. While EU payments often constitute important sources of revenue for West African governments, the security of fishing resources, vital to so many aspects of these countries, are often threatened.

The case of one West African country that signed an access agreement with the EU, Senegal, illustrates how subsidies and trade agreements can lead to threats on food and livelihood security. The Senegalese government initially implemented subsidies as a way to encourage fish production to supply domestic consumers with a protein source; fish accounts for 75 percent of the population's protein needs. But the use of subsidies has since become incompatible with this goal since the end result has been a re-orientation towards encouraging the exportation of fisheries resources and agreements with the EU for access to Senegalese fisheries. The combination of domestic export subsidies, which lower the costs of production for those fishers who chose to export, and trade advantages granted under the Lomé Agreement of 1982, which authorized customs exemptions to products originating in African, Caribbean, and Pacific (ACP) countries and increased the competitiveness of Senegalese piscatorial products in the European market, have encouraged small-scale fishers in the country to both intensify their efforts in response to international demand as well as orient themselves towards the export market.

These developments in subsidy policy, along with several fisheries agreements that have been signed with Japan and the EU, have threatened livelihood and food security in Senegal. The impact of growth in subsidy-driven exports has lead to a "breakdown in the supply of cheap protein to the population," since local producers began to switch their activities toward the capture of species of high market value in European economies rather than those affordable to local consumers. The increasing scarcity of fisheries resources has also affected the traditional, artisanal fishers, whose techniques, which require low technology and employ many people, are beneficial for local resource and employment security. Since fish have become scarcer, foreign fishing fleets have become a larger presence, and local fishers lack the equipment and fuel to travel farther out to sea where the remaining catches are located, they have been forced to make deals with European or Asian boats to remain employed. In return, locals allow the foreign boats access to areas originally reserved for their needs and use. This trend threatens local control of the resource as well as livelihood security. In a country where fish provides a vital source of sustenance for the local population, in terms of both food and employment, the way in which subsidies encourage a reorientation of resources to foreign markets is a matter of much concern for the sustainable development of the country.

A ROLE FOR THE WTO IN ACHIEVING SUSTAINABLE DEVELOPMENT OBJECTIVES

The Environmentalist-Free-Trader Alliance

As the potential threats perverse subsidies pose to the world's fisheries and those who depend on them become increasingly evident, many in the international community are seeking ways to link reforms in subsidies to those clearly also needed in management and resource use. One option that has recently become popular among environmental and sustainable development organizations and several fish-producing nations is the pursuit of subsidy reform through the WTO. This endorsement is significant in several ways. First, the connection between the liberalization of trade through the WTO has not traditionally been a cause of environmental organizations. As Grynberg and Tsamenyi note, "What is peculiar is that environmentalists, who for so long focused on economic and population growth as the cause of resource depletion, should now enter into alliance with the free traders and focus on a symptom rather than what they have traditionally seen as the root cause of the problem." Environmentalists have also conventionally highlighted the numerous potential conflicts between the liberalization of international trade and the use of the world's resources. In the context of fishery reform, however, international conservation organizations such as the WWF and the IUCN have expressed that their organizations hold several aims parallel to those of the WTO. A related second shift this "alliance" illustrates is the progress of thought concerning the role of the WTO and trade liberalization. Instead of seeing measures such as subsidy reduction purely as ways in which to influence economic factors such as market access and prices, these same measures are now being seen as capable of influencing areas of sustainable development such as the environment and human and national development.

Since its creation in 1995, the WTO has included sustainable development as one of its objectives along with the liberalization of international trade, but the linkages between the two objectives have not always been clear. The WTO promotes a world trading system based on the principle of comparative advantage, suggesting that, "In a trading system in which all nations remove protectionist measures and other barriers to market access, the net benefits to all should increase." According to the WTO, subsidies are trade-distorting measures that impede the operation of the principle of comparative advantage. What the WTO, a forum for negotiation of trade measures, cannot assure is to whom the economic rents from this process go or how resources are used in the process of trade and production.

Background on the Place of Fisheries in the WTO

Fisheries, their trade, and the subsidization of their production have yet to find their places in the WTO. Due to conflicting interests during the Uruguay Round negotiations, the issue of fisheries subsidies was not resolved at the Organization's inception, and fisheries subsidies were included under the Agreement on Subsidies and Countervailing

Measures (SCM). This agreement covers all goods except agriculture and attempts to place restrictions on the power of WTO members to provide subsidies to their industries. As UNEP reports, since 1995 a "significant change in the political context of the issue has taken place. The impetus for inclusion of fisheries subsidies on the international trade agenda has increased substantially, primarily because it has been treated not only as an issue of efficiency and equity in international trade but also as an issue of protecting natural resources from depletion." Momentum for getting fishing subsidies put on the trade agenda increased in the late 1990s, in particular at a special session of the General Council of the WTO where Iceland, with the support of nearly 20 other countries, proposed that members agree to eliminate those subsidies that contribute to overcapacity. The desire among many WTO members to climinate capacity-inducing fishing subsidies set the stage for an increased role for fisheries in future WTO negotiations.

In anticipation of the fifth Ministerial Conference in Cancun, Mexico of September 2003, a group of states calling themselves "The Friends of the Fish" lead the attack on fishing subsidies and their apparent threats to world trade and sustainable development. The group, which formed prior to the Seattle Ministerial Conference of 1999, successfully campaigned for inclusion of a paragraph in the Conference's draft declaration but failed to see an agreement be reached as the meeting broke up without a final ministerial declaration. Australia, Chile, Ecuador, Iceland, New Zealand, Peru, the Philippines, and the US are the leaders of the "Friends." Japan and South Korea, which have large subsidy programs for their fishing fleets, have argued that fishing subsidies should not be treated in terms of having trade-distorting effects and question the linkages others claims the subsidies have with resource exploitation.

The EU is between the two sides; traditionally having sided with Japan and South Korea, it has recently been less defensive of its subsidies. Hoping for better results in the Cancun negotiations of 2003, the "Friends" then submitted a proposal to the WTO in April of 2002 outlining the "potentially harmful trade, developmental and environmental effects of subsidies to the fisheries sector" and suggesting the need to address these effects. While negotiations at Cancun also failed to deliver a consensus on the issue, the group and their allies continue to push for reforms through the WTO that would eliminate what they consider perverse subsidies in the sector.

Input of International Organizations

The call for the reform of fishing subsidies has come not only from member states of the WTO, but also from international non-governmental organizations. Economic organizations that have become involved include the OECD, which undertook the first study to measure fisheries subsidies in OECD countries in the early 1990s, and the Asia-Pacific Economic Cooperation Forum (APEC), which collected data on fisheries sector support programs in member governments. The FAO has played a crucial role in putting fishing subsidies on the radar; by their account they "sounded the alarm about the negative effects that subsidies were having on capture fisheries" in 1992, which "led governments of major fishing nations and various Intergovernmental Organizations to look into the issue." These organizations stress the link between fishing subsidies and their ability to cause perverse economic incentives and overcapacity.

Organizations with conservation and sustainable development agendas have also added their voices to the criticisms of fishing subsidies, and many have iterated a possible role for the WTO in the attack. The WWF, for example, has published much literature emphasizes the link between government subsidy payments and overfishing and, in one essay, suggests that "The World Trade Organization (WTO) today has an unprecedented opportunity to help improve the environmental and economic health of the world's oceans by disciplining harmful fishing subsidies." The IUCN has also described a role for subsidy reform and the WTO in the achievement of fishery use within sustainable development aims. One publication examines possible "synergies" between "international trade, sustainable fisheries, the environment, and sustainable development." By focusing on the causal factor of subsidies in overfishing and resource depletion, these organizations have created a bridge between a trade measure and the state of natural resources, thereby creating a role for an economic organization such as the WTO in the pursuit of sustainable development aims.

The Sustainable Development Concept

The context into which governments and organizations are currently trying to put subsidy reform, that of sustainable development, pulls together concerns that are a combination of economic, developmental, social, and environmental. The concept of sustainable development became prominent at the 1992 United Nations Conference on Environment and Development, which defined the idea as "meet[ing] the needs of the present without compromising the ability of future generations to meet their own needs." With its incorporation of a commitment to poverty eradication, reductions in disparities of standards of living, and environmental protection as an integral part of the development process, sustainable development "challenges the system to reconcile trade objectives with broader environmental, social, and cultural imperatives, and also raises concerns as to the issue of what constituted effective and legitimate development, rather than measuring development in narrow terms of economic growth." Sustainable development in terms of fisheries includes meeting the economic, social, and cultural needs of coastal communities, addressing the concerns of developing countries to maintain revenues from trade that are necessary for development, monitoring the health of fish stocks, and surveying the surrounding situation of biodiversity. This perspective encompasses much more than either the strictly economic side-benefits to increased fish production and trade, or the conservation side-sizes of fish populations.

Reform of fishing subsidies, an issue that seems to have become a nexus of economic, environmental, developmental, and social interests, is defined by many as a potential 'win-win-win' situation for all concerned. According to the WFF, "the WTO and its members have been promising for years to negotiate new rules that deliver true 'win-win-win' outcomes for trade, the environment, and sustainable development," and the fisheries subsidy issue presents the "first real" opportunity to move toward this objective. In suggesting that such a situation exits, 'win-win-win' outcome proponents conjecture that the same measure, here the elimination of perverse fishing subsidies, could deliver favourable benefits to trade by increasing market access for producers, the environment by reducing the capacity of global fishing fleets, and sustainable development by assuring resource and livelihood security to those in developing countries.

The Argument for the Use of Subsidy Reform in the WTO to Achieve Sustainable Development Objectives

The argument for proposed changes brought through the WTO, that many view as having the potential to deliver this 'win-win-win' outcome, are based on three main justifications of subsidy elimination: trade, sustainable development, and the environment. The trade foundation of the argument deals with the way in which cost-decreasing, rent-increasing subsidies may deflate prices on the international trading market because of lower costs for producers due to government support that absorbs a portion of costs. The submission made by the "Friends of the Fish" to the WTO stated that subsidies in the sector caused distortions to trade through their "effects on prices and production levels," and that the transfers "significantly distort trade in fisheries products by lowering costs of subsidized production, increasing the volume of subsidized production and thus distorting competitive relationships in markets for fisheries products." Concerns for the impact of such distortions are also voiced by organizations, mainly in the context of how they might adversely affect the economies of developing countries.

An IUCN report suggests that, for many developing nations, "trade liberalization offers possibilities for addressing declining terms of trade," and that efforts to reduce subsidies could have the positive impact of enhancing " export opportunities for countries currently shut out of export markets due to subsidized competition." These factors are important for developing countries in particular because of questions of equity in trade; as Charlie Arden Clark of the UNEP expressed, those best able to subsidize themselves (wealthy developed countries) will best be able to obtain the resource and trade it on the world market.

The importance of fishery product exports to the economies of many developing countries is also an important part of the argument; the FAO highlights how "exports of fishery products are vital to the Least Developed Food Deficit Countries, in terms of export revenue, employment and income generation." The potential benefits to elimination of perverse subsidies in the fisheries sector such as increased market access for non-subsidized competitors comprise part of the argument in how subsidy reform might form an important part of sustainable development.

The potential positive impacts of addressing fishing subsidies in the context of an international trading system could extend beyond those that are purely economic, though, and contribute to several important aspects of human and national development. Due to the way in which they can threaten food and livelihood security, the effects of access agreements and the foreign fleets they bring into the waters of developing countries are being targeted in particular as a way to move toward development aims. An IUCN report notes that one positive impact might be to "Provide greater opportunities for domestic fisheries industries in developing countries by reducing subsidies to foreign distant water fleets." If access agreements were eliminated, there may be less competitive pressure from foreign industrial fleets, which "effectively compete for fish with local artisanal and industrial fisherman, thereby threatening their livelihood security, and undermine the economic development of poorer West African nations."

The presence of foreign fleets and the governmental revenue they bring are also of little development assistance to local fishing communities. As Aimee Gonzales of the WWF explained, the fees that developing country governments receive as access payments are not used to access capacity or monitor stocks, measures that might enhance local capacity to develop a sustainable fishing sector. She also described how there is little "trickle-down" effect from the presence of the industrial fleets in the coastal communities; since the big, foreign boats have the ability to process their catches on-board rather than use local processing plants, they contribute little to the local economy. The elimination of access agreements may, in some cases, also help increase food security by increasing the share of fish that go to foreign fleets for export. Developing countries hypothetically stand to gain much from a decreased presence of foreign fleets in their waters.

These potential gains for development in eliminating foreign access to fisheries of developing countries have inspired proposals from several organizations for reform. An IUCN report, for example, challenges the agreements to be considered subsidies under the WTO SCM Agreement, as well as proposes a new international agreement that "includes provisions prohibiting the payment by governments of any part of the costs of access to fishery resources." The WWF also recommends that the WTO expand its definition of "subsidy" to include the access agreements, and to address the issue of sustainability by "considering whether subsidies contribute to excess capacity, over-fishing, or unsustainable fishing practices." Under such propositions, a subsidy may be prohibited under WTO rules, even if it does not cause trade-distortions, as long as it is seen as being counter-active to sustainable development goals.

The environmental impacts of cost-reducing, rent-increasing subsidies are looked at mostly in terms of how the subsidies contribute to fishing fleet capacity. The link between subsidies and overcapacity is crucial in understanding both the role subsidies have played in causing the current state of resource depletion and what role they might play in the future in the pursuit of more sustainable fishing behaviour. According to an UNEP study on the effects of subsidies in the fishing sector, "The key issue for analysis and assessment in regard to the linkage between subsidies and overfishing is whether and to what extent fisheries subsidies contribute to fleet overcapacity." Fishing fleet overcapacity has been identified as a major threat to fishery resources, and global fishing fleets are currently estimated to be up to 250 percent greater than needed to catch what the world's oceans can sustainably produce. Carl Gustaf Lundin, Head of the Marine Programme at the IUCN, emphasized the need to reduce capacity in order to achieve fisheries resource security.

Because of the threats they pose to the environment, governments and organizations suggest that capacity-inducing subsidies be prohibited by the WTO. An IUCN report proposes that various modifications be made to the WTO SCM Agreement to "ensure that all capacity inducing subsidies are covered by the agreement." The WWF and the "Friends of the Fish" also suggest that the WTO rules address the issue of subsidies that contribute to excess capacity and contribute to fish stock depletion. These groups hope that, if subsidies that increase fishing fleet capacity are targeted and eliminated, fish stocks may be able to recover in the face of decreased fishing intensity.

While most of the argument for the link between the elimination of perverse subsidies and certain sustainable development goals has been based on these three objectives, some feel that a role for the WTO may in itself add to progress. They believe that the WTO can potentially increase political action in the area of fisheries and improve other important aspects that need reform, such as fisheries management. Aimee Gonzales at the WWF explained that, since subsidies are where the politics are, action through the WTO might spur "parallel" action among other bodies to create good management regimes and, thus, sustainability. Hugo Cameron of the International Center for Trade and Sustainable Development (ICTSD) agreed with this idea, noting that the fact that the subsidies issue is being brought up at the WTO will give fisheries "air time." He also felt that recent improvements in the EU's Common Fisheries Policy were due in part to pressure from governments and environmental groups for reform inline with WTO rules. The "Friends of the Fish" submission to the WTO describes a role for the WTO in the context of a situation where action is clearly needed: While many factors contribute to depletion of the world's fish stocks, harmful subsidies are one important factor within the WTO's competence to address....As the forum in which global trade rules are negotiated and enforced, the WTO has a unique and important role to play as part of a multi-layered global response. The WTO's mandate, experience, expertise, and pre-existing framework of rules and dispute settlement mechanisms are critical to dealing effectively with the trade-related aspects of this problem. An integral justification of the proposed role for subsidy reform through the WTO has become the benefits the WTO can uniquely contribute to the situation due to its international political importance and prominence.

WTO Subsidy Reform as a Method for Achieving Sustainable Development

Given the complex and far-reaching impacts organizations and states conjecture for fishing subsidy elimination through the WTO, it will be difficult for actual reforms to live up to their model. Certain benefits seem assured. On a net basis, trade liberalization can increase the economic gains of a developing country by increasing their market access. And, in a vacuum, the elimination of subsidies can decrease productive capacity and leave more resources available for the use of those who were formerly excluded. Sustainable development, though, is a much taller order. It means assuring that certain aims can be met in the context of forces such as trade and resource use. Whether or not WTO action on fishing subsidies can achieve these aims is the question. The ability of subsidy elimination to decrease overfishing by subsidized fleets or create market access where there perhaps was none before is not under debate-what are are the implications and possible benefits of this action.

Regarding the environmental sphere of the argument-the use of subsidy reform for conservation aims-the security of fisheries resources has many factors beyond that of the subsidization of fleets. While the theoretical reasons that support the role for cost-reducing, rent-increasing subsidies as a potential cause of increases in capacity are strong, and many empirical cases have illustrated instances when increases in such subsidies played a pivotal role in overcapacity, the extent to which they have done this is still under debate. The importance of management in the relationship between subsidization and overfishing has been found to be incredibly significant since effective management may alter the impact of subsidies on resource use. If fisheries are well managed and impose regulations and/or quotas, additions to capacity may not result in additional effort. The OECD Committee for Fisheries reports that the effects of subsidies on fisheries resources are theoretically different depending on the type of management regime present. As could be expected, the effects of subsidies on open-access-type management regimes include overexploitation and increased capital and labour being attracted to the industry. In contrast, under an effective management scheme, the same subsidies had no effect on resource sustainability. By conjecture, if effective management schemes were also present in developing countries, the degradation of their resources in the face of highly subsidized fleets may not seem so imminent. Theoretically perverse subsidies clearly create overcapacity and resource degradation in many cases, but do not stand alone as the causal factor of current fisheries depletion.

The validation for subsidy elimination based on trade illustrates the complexity of achieving a true 'win-win-win' situation, especially when the interests of one of the main supporters of this view, the "Friends of the Fish," are examined. The group, described by Hugo Cameron at the ICTSD as a group of relatively competitive fishing countries, is lead by the US and New Zealand, fishing nations that subsidize their fleets relatively little. Support by these two countries in particular seems to be rooted in their economic concerns over the competitiveness of their fishing fleets. The US government apparently blames the lack of profitability of its fleets on the low world fish market prices, which it sees as being held down by the subsidies in distant water fishing nations. Eliminating subsidies in the EU and Japan, for example, would theoretically increase prices and improve the US's market share. According to Grynberg and Tsamenti, environmentalists who support trade liberalization as a means to sustainable development and environmental security are in effect supporting US trade interests. To the "Friends of the Fish," eliminating subsidies in the fishing sector is a way to increase the competitiveness and productivity of their fleets.

What the motives of the US and the "Friends" highlight is that there is a significant endorsement of free market principles in the attack on fishing subsidies, principles that, if implemented, might unleash threats on areas crucial to the pursuit of sustainable development and resource security. An IUCN report suggests the possibility that "the free functioning of comparative advantage in trade relations may not per se encourage the best long-term allocation of resources in the fisheries sector." This possibility becomes especially acute when the lack of effective management of developing country fisheries is recognized. While, in many industrialized countries stocks are monitored and assessed and few fisheries remain purely "open access," developing countries lack the capacity and funds to effectively manage their fisheries. The WFF reports, for instance, that "Traditional management methods like surveillance patrols, ecological monitoring, or measures to address pollution from both land and sea have met with only limited success because they are simply too costly for the region's cash-strapped institutions." This asymmetry in management between the North and the South may actually encourage fishing in developing country waters, where management is more lax.

This idea is supported by economic theory, which warns of a comparative advantage that can seem to be present for the production of common property resources in the South due to their less well-defined management or property rights schemes. As described by the economist Graciela Chichilnisky, "lack of property rights alone can create trade" between two countries that are otherwise identical since the ill-defined property rights in one country alone will cause the overuse of a resource in that country. Fisheries biologist David Boyer explained that increased market access for developing countries may indeed increase pressure on fisheries as fishers in these countries scale-up their fleets in response to international demand for high-valued fish. In his opinion, market liberalization may well "shift pressure to greater fishing for trade from countries that have little environmental protection and the inability to enforce international treaties or their own national laws."

He also notes that, without "deliberate domestic policies to facilitate restructuring" in the face of trade liberalization, "liberalization may, at least in the short and medium term, actually work against growth, employment, poverty alleviation, and other components of sustainable development." Liberalization of fish trade, one of the impacts of subsidy reform, may only continue the current pattern of resource overuse that threatens the development and resource security of developing coastal nations, or, at worst, increase this overuse.

Better management of fisheries in developing countries increases the chances of "winning" on several development fronts since it provides some assurance of how the resource would be used after the elimination of subsidies in addition to "less," or "more in line with market principles," as the subsidy elimination argument suggests. As described, free trade of fisheries resources may put pressure on the resource and do little to assure livelihood security for local communities. Effective management schemes could enforce quotas and regulations, assuring livelihood security for future generations. Current livelihood security in many developing fishing nations further depends on the lessening of competitive pressure from distant-water fishing fleets from developed nations.

Subsidy elimination would very likely lessen the number of EU or Japanese boats in these nations' waters, but this change says little about which boats, such as those of Iceland or Argentina, known to be very efficient, might enter next. Even when subsidies are removed, high global demand for fish will still exist, as will the reality that a large percentage of the world's fish are located in the waters of Western Africa and

Western South America. In my opinion, management is one of the only likely mechanisms that can relieve such market pressures on the resource. Food security may be improved little by removing subsidies if, as the trade scenario described suggests, producers still have large incentives to focus their efforts on high-value fish for export to developed countries. If management were effective, though, catches by industrial or foreign fleets could be limited and some resources could be reserved for local population use, a method that has been proposed in Mauritania. Management of fisheries in developing countries is not only a key aspect of resource sustainability, but also a necessary guard against potential threats of subsidy elimination.

Organizations and the "Friends of the Fish" argue that certain characteristics of the WTO, such as its prominence in political and economic arenas, may help sustainable development aims by leading to a chain reaction of reforms in other areas. This argument's merits aside, the international scale of the WTO, whose rules "are intended to help ensure that trade flows as smoothly, predictably, and freely as possible" at an international scale, poses potential threats to interactions at other levels. For example, as the push for action on fisheries issues at the WTO increases, those engaged in fisheries agreements at the regional level are raising concerns about potential conflicts between regional agreements and international WTO rulings. Regional Fisheries Organizations (RFOs) adopt their own trade measures between states in a region in order to achieve purposes such as compliance with conservation and management regimes.

RFOs can ensure cooperation through requirements for documentation, prohibiting landings from non-compliance vessels, or implementing trade measures. Given their aims for effective management, RFOs present an opportunity for reform in an arena crucial to fisheries resource conservation and largely left out of the impacts of subsidy reform. But, as Richard Tarasofsky forewarns, RFOs, which "seek to discriminate in favour of goods produced in accordance with their conservation and management aims, are potentially in conflict with these

[WTO] provisions," which mandate equal trade treatment of like products. A winning situation at the WTO may be a loss at the regional level if WTO rules that assure the smooth international trade of equally treated products compromise the ability of regional organizations to institute effective management schemes and assure resource conservation.

The threats the international focus of the WTO pose for RFOs are an example of what action on fisheries subsidies at the WTO could mean for developing country national governance. Governance of fish stocks currently rests at the national level-since the UN Law of the Sea Convention of 1982 states have had sovereignty over stocks that lie within 200 nautical miles of their shores. As the subsidy reform issue highlights, management is a key aspect of development and resource security in developing coastal nations. National governance is thus an unavoidable part of a solution and holds many opportunities for conservation of fisheries resources by intervening in the domestic situation and creating management solutions. The government of Senegal, for example, announced in September 2003 the creation of four marine protected areas that are of vital importance to local fishers and the domestic economy.

While this action was hardly a subsidy, at the level of the WTO some states may argue that any intervention by a government in the fish sector destroys the market. Hugo Cameron of the ICTSD gave a current example of this argument when he explained how, although many states are trying to use subsidies to reduce capacity by using subsidy funds to buy-back vessels or employ people to repair ports, states such as those in the "Friends of the Fish" may argue that this still provides people an incentive to remain in the industry and distorts the market. As Carl Gustav Lundin warned, it can be a "slippery slope" when international bodies limit the interventions of national governments. In order to achieve sustainable development objectives, WTO rulings on fisheries must not interfere with developing nations abilities to use government support measures in their domestic economies and resource management sectors.

Although governance of fish stocks rests at the national level, sustainable development objectives require that livelihood and food security be met at the local level, and that future solutions do not stop at questions of national access or resource use. Conservation, fair economic activity, and development are meant to be non-national goals, even though, as the fisheries subsidies issue highlights, they have often been dealt with in terms of national policy.

Ending the analysis at the national level, though, is a fault. Current threats in the fisheries sector at the local level come from the way in which trade liberalization through the WTO may not assure that the distribution of benefits from export revenues correlate to any measure of development objectives. Increases in national export revenue do not necessarily represent changes in people's access to resources for food or livelihood.

Instead, they may simply reinforce the existing economic structure within the community. This is especially so in this case due to the lack of effective management of fisheries resources, which, as described, may only add to resource depletion under trade liberalization. Given the way the WTO does not monitor such local dimensions, it is incapable of assuring any outcome outside that which relates to the state as a single economic entity. A real resolution to the crisis would take the inner view of the national situation-most importantly here the level of resource management-into account and tailor policy solutions to this.

CONCLUSION

In order to achieve a true 'win-win-win' solution, subsidy reform must deliver not only benefits on the international level; global levelling of trade advantages in the economic sphere, net conservation of resources in the environmental sphere, and a targeting of resources toward development aims, but gains at all levels. This is perhaps the most challenging aspect of the role proposed for the WTO, a fundamentally international organization. In order to be successful, it needs to prove that it can improve the lives of people down to the local level. The concepts of sustainable development are an entry point into such progress, but, as this paper describes, there are serious limitations to relying on an international economic institution to deliver such goals. The threats which have been outlined; the lack of assurance that gains at the national level through increased export revenues and trade liberalization will deliver resource or livelihood security, the threat to national autonomy due to WTO intervention in the responses of state governments to the fisheries crisis, and the dire need for effective management in the fisheries of developing countries, represent needs on which WTO action as it is presently seen cannot fully deliver. Only when proposed reforms can assure that objectives are met in the three pillars of sustainable development-environment, trade, and development-as well as at all levels of organization and governance can any outcome be considered to be winning.

4

Sustainable Development

In fisheries, development may be defined as "a process of change through which sustainable and equitable improvements are made to the quality of life for all or most members of a society" (Bailey and Jentoft, 1990). Adding the need for sustainability implies that these improvements need to be achieved without risk to the long-term stability of the ecosystem concerned.

In 1987, the World Commission on Environment and Development defined sustainable development simply as "Development that meets the needs of the present without compromising the ability of future generations to meet their own needs." This definition recognises that human needs can change with time, and that preservation of the integrity of the environment is necessary to ensure that future generations can realise the full potential of their own development needs. Another, more elaborate definition states that "Sustainable development is the management and conservation of the natural resource base, and the orientation of technological and institutional change in such a manner to ensure the attainment and continued satisfaction of human needs for present and future generations.

Such development conserves land, water, plant and genetic resources, is environmentally non-degrading, technologically appropriate, economically viable and socially acceptable." This definition was developed by the FAO Committee on Fisheries in 1991, and acknowledges the need to achieve a workable balance between guaranteeing satisfaction of both present and future human needs including fulfilment of social and economic demands, and conserving the natural resource base. In response to the general goals identified, FAO developed a Code of Conduct for Responsible Fisheries (FAO, 1995) that provides principles for ensuring sustainable exploitation of marine resources.

SEARCHING FOR OPTIMUM FISHERIES MANAGEMENT

The 1991 FAO definition noted earlier recognises the need for management of the resources concerned as one of the essential actions for achieving sustainable development. To accommodate the other components of the FAO definition, it is clear that such management must embrace a range of objectives, incorporating biological, economic, and social elements. Knowledge of the status of the resources and the likely impact of fishing activities and other factors on these resources provide for informed and successful management of fisheries.

In the early days of fisheries management, scientists focused on the biological modelling of fish stocks (Graham, 1935; Beverton and Holt, 1957). Assuming that fish stocks respond predictably to moderate levels of exploitation, with a defined equilibrium state, fisheries scientists are able to calculate the maximum level of catch that would allow sustainable exploitation. This sustainable maximum catch is referred to as the Maximum Sustainable Yield (MSY), and continues to be a key reference point in present-day fisheries management (Gulland, 1983).

MSY takes into account only the sustainability of the resource and so satisfies only a biological objective. In view of this, fisheries economists have argued that fishing is conducted as a business and that the economic benefits of the activity must not be ignored.

This led to the development of another management reference point, referred to as Maximum Economic Yield (MEY). MEY is calculated based on optimising the difference between the cost of fishing and the income gained, and usually occurs at a lower fishing effort than the fishing effort occurring at MSY (Hersoug, 1996).

Subsequent to the development of MEY, social scientists have noted the importance of addressing social concerns, to ensure more equitable distribution of economic benefits. The social objective of Maximum Social Yield (MsocY) was then recognised, and the associated reference point varies with the fishery situation.

These biological, economic and social management objectives, as currently defined, are now facing yet another challenge, with the development of chaos theory (Smith, 1990 in Symes, 1996; Wilson and Kleban, 1992 in Symes, 1996). Chaos theory suggests that although nature is non-random, it is unpredictable. The equilibrium state of fish populations assumed by MSY is not accepted, and fish populations are believed to vary unpredictably within limits (Symes, 1996). Chaos theory therefore proposes 'ecologically adapted management', based on knowledge of the long-term stable ecological relationships, and requires more flexible management systems.

ACHIEVEMENT OF RESPONSIBLE FISHING

Sustainable fisheries development can be achieved through responsible fishing, which considers rational fishery management objectives that address a range of issues including the status of the resource, the health of the environment, post-harvest technology and trade, as well as other economic concerns, social benefits, legal and administrative support. In the case of shared resources, a co-ordinated approach to responsible fisheries management is essential, and Caddy and Griffiths (1995) proposed the following actions:

Regulate Fishing Effort

It is crucial to control fishing effort, and to avoid financial incentives that would contribute to excess fishing capacity. Excess fishing capacity and overcapitalization threaten the sustainability of the resource, as well as the industry.

Establish Code of Conduct for Responsible Fishing to Guide Management Plan

This is needed to maximize benefits from the fishery, while avoiding wastage caused by indiscriminate fishing practices. In particular, it is important to reduce catches of undersized fish and non-target species, and to avoid use of gears that have a negative impact on the environment. Establish and support regional/international fishery commissions and organisations concerned with management of shared resources

The successful management of shared resources requires effective coordination by all the countries concerned. The relevant international fisheries agreements promote participation in, and financial support for, the work of these commissions and organisations.

Regular Consultation among Harvesting Countries

Parties sharing the resources need to consult and collaborate regularly so as to promote understanding and full cooperation. Set agreed management objectives and related reference points, incorporating a precautionary approach Agreement on management reference points during the early stages of the fishery will help to ensure full cooperation of participants with management decisions. Where there is scientific uncertainty, a precautionary approach to management is recommended.

Develop Contingency Plans

Management plans should incorporate some contingency for dealing with sudden and unpredictable environmental changes caused by man-made or natural disasters.

Develop Mechanism for Resolving user Conflicts

Management should provide mechanisms for handling problems arising from resource user conflicts.

Protect Biodiversity

The biodiversity or species richness of an ecosystem is an important measure of ecosystem health. The preservation of biodiversity will ensure that present human development activities do not threaten the ability of future generations to meet their own needs.

Protect the Environment

There should be monitoring and control of waste disposal and pollution. In addition, every effort should be made to prevent discarding of entangling material that could trap and kill species or physically damage the environment.

Promotion of Research

Research should be conducted to support and inform various aspects of management.

Optimise Social and Economic Stability

There should be fair and equitable distribution of benefits derived from the fishery.

Constraints to Caribbean Island States

For many small developing island states within the Caribbean, there are a number of constraints that pose significant challenges to the sustainable development of fisheries, including the following:

- *Fishing Practices:* Destructive fishing methods such as dynamiting fish are still practiced in certain areas. Additionally, fishers continue to use illegal mesh sizes that catch very small fish, and non-target species that may have little or zero market value.
- Inadequate knowledge of the resource and ecosystem

  Many Caribbean islands do not have sufficient detailed data to permit an accurate evaluation of the status of their resources and a good understanding of their marine ecosystem.
- Inadequate knowledge and recognition of social and economic conditions

  Many governments within the region still do not have an adequate appreciation of the social and economic potential of sustainable fisheries development, and hence still invest minimum resources in fisheries development and management activities.

- *Absence of Long-term Policies:* The lack of long-term policies makes it difficult for Caribbean countries to maintain pace with rapidly evolving trends in global fisheries management approaches and trade strategies.
- *Capacity:* In many Caribbean countries, fisheries administrations are poorly staffed, and there are limited numbers of skilled and knowledgeable fisheries technicians, scientists and managers, and limited equipment and funds available for basic tasks such as data collection and research. There are few research institutions and regional organisations within the region, and some of these suffer similar limitations in available funding and a broad range of technical expertise.
- *Resource user Input:* In the Caribbean, fisherfolk organisations have not actively participated in the management process, mainly owing to a lack of good organisational and administrative skills. On the part of the fishers, this has resulted in a lack of trust in, and respect for, the governments that are responsible for making fisheries management decisions.
- Capability for monitoring, control and surveillance

  In many instances, there is little or no capacity for monitoring, control and surveillance activities, and no associated legislative framework.
- *Post Harvest Aspects:* Caribbean countries are just beginning to develop their export markets for fish and fish products, and face the challenge of satisfying stringent standards in all aspects of fish handling, processing and packaging, recently established by the importing countries.
- *Facilities:* Equipment and facilities are not upgraded regularly to deal with expanding fishing activities and increasing management demands.

- *Habitat Degradation and Pollution:* For some time, there have been signs that the Caribbean ecosystem is under stress (Richards, and Bohnsack, 1990). Degradation of habitat, particularly in the coastal areas has resulted from uncontrolled coastal development activities mostly associated with expanding tourist industries and overpopulation problems, as well as extensive sand mining and deforestation.
- *Financial Resources:* Many Caribbean countries are small island developing States with limited financial resources. This impacts negatively on their ability to cope with the wide range of issues required for successful sustainable fisheries development.

Caribbean governments need to appreciate the social, economic and financial potential of sustainable fisheries development. Fisheries administrations need to be given more financial resources, and properly staffed and equipped to address effectively and completely all aspects of sustainable fisheries development and management. Additionally, small island states should recognize the benefit of, and work towards, regional coordination in fisheries management activities, including sharing of expertise and resources for education, research, technology, monitoring, control and surveillance activities, and development of the relevant legislative framework.

Having removed constraints to responsible fishing, fisheries management must then strive to achieve a workable balance between defined objectives protecting the resource and its environment, and those seeking fair and equitable distribution of viable economic benefits.

CURRENT PROBLEMS IN THE MANAGEMENT OF MARINE FISHERIES

The public perception of fisheries is that they are in crisis and have been for some time. Numerous scientific and popular articles have pointed to the failures of fisheries management that have caused this crisis. These are widely accepted to be overcapacity in fishing fleets, a failure to take the ecosystem effects of fishing into account, and a failure to enforce unpalatable but necessary reductions in fishing effort on fishing fleets and communities. However, the claims of some analysts that there is an inevitable decline in the status of fisheries is, we believe, incorrect. There have been successes in fisheries management, and we argue that the tools for appropriate management exist. Unfortunately, they have not been implemented widely. Our analysis suggests that management authorities need to develop legally enforceable and tested harvest strategies, coupled with appropriate rights-based incentives to the fishing community, for the future of fisheries to be better than their past.

The United Nations Food and Agriculture Organization (FAO), which monitors the state of world fisheries, has estimated that since 1990 approximately one-quarter of fish stocks have been overexploited, depleted, or are recovering from depletion (17%, 7%, and 1%, respectively), with the Northeast and Northwest Atlantic, the Mediterranean, and the Black Sea being the areas with the largest number of depleted stocks. Many authors have elaborated on these conclusions, documenting the poor state of fisheries worldwide. Nevertheless, the situation, although serious, is not catastrophic, and there are grounds for optimism. There have been successes of fisheries management, and there is an understanding of what is involved in successful fisheries management and of the requirements for its implementation. These issues are explored in this review.

The management of commercial fisheries clearly requires a good scientific understanding of the behavior of the exploited stock or stocks. The science that is used to assess commercially exploited species is still dominated by the population models developed by Beverton and Holt for single-species assessments some 50 years ago. The availability of substantial computing power has meant that sophisticated estimation methods can be used, and an appreciation of the way in which fish stocks respond to environmental variability is readily incorporated in scientific advice. Calls for a more ecosystem-orientated approach have been voiced for some while, but the paucity of data and the demands of multiparameterized multispecies models means that most ecosystem considerations in practical stock assessment tend to be ad hoc manipulations of the single-species approach.

What has developed is a realization that effective management requires an understanding of how the fishery system is performing relative to reference points. The most commonly used reference points are those relating to the size of the stock itself and the fishing mortality that will result in these stock sizes, given existing relationships between the stock, recruitment, natural mortality, and growth. A typical "target reference point" is the biomass necessary to produce maximum sustainable yield (BMSY). However, such targets do not explicitly recognize threats to the stock. To address this issue, stock size "limit reference points" are usually defined or interpreted as the stock biomass below which recruitment becomes substantially reduced. Clearly, it is important to avoid situations where the stock is at or below this level. Accordingly, management should aim to have as a target a level of stock size that carries a low risk (allowing for scientific uncertainty) of the stock dropping below the limit reference point. This could mean having a target level of fishing mortality that provides stock sizes above BMSY.

UNDERSTANDING FISHERIES MANAGEMENT

Competent scientific advice based on appropriate data is far from ubiquitous in the fisheries world, and even in ideal situations, fisheries management has often been unsuccessful. The success of a management system is often defined in terms of biological, economic, social, and political objectives. Clearly, economic and social objectives will not be met while a stock is in such a depleted state that the long-term sustainability of the fishery is threatened, but equally, biological objectives are unlikely to be met without consideration being given to economic and social objectives.

Hence, we argue that an understanding of the fishery management process can only come from analyzing the capacity and incentives of the two key stakeholders: the fishing community and the management authority. This is not to belittle the importance of other stakeholders, such as recreational fishers and environmental groups, who have important roles in the management of certain fisheries.

Where management is weak or nonexistent, the economic factors underlying overfishing in commercial fisheries have been generally understood since the 1950s. In short, when multiple fishers compete to catch fish from a given population, each fisher maximizes his net income by continuing to fish as long as the value of his catch exceeds the cost of catching it. An equilibrium, called the bionomic equilibrium, is reached only when fishing has reduced the fish population to a level at which catch rates are barely sufficient to cover the costs of fishing. The population is then maintained at this level through biological processes of natural growth and reproduction. Thus, if the price:cost ratio is high, the bionomic equilibrium will result in a low stock of fish, and hence a low annual catch level; two characteristic features of overfishing. In addition, the so-called economic rents (total revenue minus total costs) from the fishery will equilibrate at zero, resulting in minimal overall economic efficiency.

Many management authorities seek to meet their objectives by setting output controls in terms of a total allowable catch (TAC) for the year and closing the fishery when the year's cumulative catch has reached the TAC. Restrictions on fishing gear, fishing season, and fishing areas, as a supplement to the TAC, may also be imposed. If the TAC is correctly specified and enforced, this method should maintain a stock level well above that of bionomic equilibrium. If, however, the TAC and the science behind it are not respected by fishermen and not adequately enforced by authorities, widespread illegal fishing can occur. A recent example is in the eastern Baltic cod fishery where illegal fishing contributes to true catches being some 35 to 40% higher than reported.

Efficient enforcement can be difficult. Put simply, fishers will be deterred from breaking fishing regulations if their expected loss from detection and successful prosecution exceed their expected gain. In many fisheries, the probability of detection of illegal activity and the penalties are not sufficiently high to act as a disincentive.

Strong management can ensure that biological targets are met, but it is essential that regulations are enforceable. and this has often proved to be difficult. Less-than-perfect enforcement can lead to illegal fishing, poor scientific data, and a failure to meet biological targets. Input measures, such as limiting the number of vessels or restricting available season length, are usually more easily enforceable than output measures such as TAC. However, control via input measures is vulnerable to effort creep, whereby operators increase the fishing power of their vessels through technical means. Nevertheless, monitoring of vessel performance over time and adjusting the allowable level of effort have allowed successful effort control to be implemented.

Overcapacity

Simplistically, it would seem that positive economic rents should also emerge in a TAC-regulated fishery. In reality, many TAC-regulated fisheries have experienced an unexpected increase in fishing capacity, as additional vessels enter the fishery in response to (temporarily) positive rents. Economic models then predict a regulated bionomic equilibrium, in which economic rents (net of fixed costs) again equilibrate at, or near, zero. This situation currently exists in many of the world's regulated fisheries; overcapacity of fishing fleets is widely perceived as a major impediment to achieving economically productive fisheries. It is thus ironic that such overcapacity is usually generated by the management system itself, although it can also result from high profitability during the initial phase of a newly developing fishery.

Overcapacity is widely recognized as a major problem affecting world fisheries. With its attendant social and economic problems, overcapacity can, via the political process, lead to the erosion of management control. It is also understood to be one of the results of subsidizing fisheries, which even today is estimated to be several tens of billion U.S. dollars per year. Such subsidies directly undermine the sustainability of fisheries because they lead to a bioeconomic equilibrium with high levels of fishing and low stock size. In several fisheries, government funds have been used to buy out excess fishing capacity. For various reasons, such buyback programs have been less effective than expected. First, often only the least efficient vessels are bought up, leaving total fishing capacity largely intact. Second, the buyback program by itself does not remove the economic incentives underlying overcapacity, which tends to increase once the buybacks are completed.

Thus, the underlying cause of the dual crisis of overfishing and overcapacity, as well as other undesirable outcomes, such as habitat destruction and incidental kills of untargeted species, can be found in the economic incentives of fishers who compete for their annual catches. These incentives are not affected by management strategies that retain the competition between fishers for a common-pool resource. Perhaps the most important development in fisheries management over the past 20 years has been the recognition of this fact and the introduction of rights-based management in several regimes. Indeed, it has been argued that of the tools at the disposal of managers, more emphasis needs to be placed on incentive-based approaches that better specify community and individual harvest or territorial rights, in additional to public research, monitoring, and effective administrative oversight.

Transferable Quotas

An alternative management strategy based on individually allocated transferable annual catch quotas (ITQs, or individual transferable quotas) is now in effect in several fishing nations, including Australia, New Zealand, Iceland, Canada, and Namibia. A well-organized rights system alters the economic incentives of fishers, who no longer compete for their catches, so that highly competitive fishing no longer takes place. The guarantee to fishers of a certain proportion of the catch allows them to make rational economic choices about where and when they catch fish. An ITQ system goes further, allowing the industry to settle on a fleet capacity that optimizes individual economic yield to vessels or cooperatives, although this of course can still be distorted by inappropriate subsidies.

In addition, ITQ fishers may often be expected to favor management actions that protect and enhance fish populations, because the value of a quota share increases as stocks become more abundant.

Problems that may arise, such as misreporting or high-grading of catches, have been successfully countered by the use of observers, required by the management system but paid for by the industry; observers are used extensively in the U.S. Pacific fisheries, Australia, and New Zealand.

Experience with ITQ systems shows that many fishers willingly support and adhere to conservative management strategies and may also avoid fishing practices that endanger habitat or threaten other species, so long as they are guaranteed long-term rights. But this does not mean that enforcement and scientific monitoring are unnecessary in ITQ systems; both are essential unless catch levels are set at precautionary low levels. It is thus unsurprising that the two countries with perhaps the most fully developed ITQ systems, New Zealand and Iceland, have some of the highest costs of management per fishing vessel.

Several authors have pointed to instances of successful fisheries management in both the developed and developing world. Among their conclusions are that incentive structure, institutional capacity, and participation of stakeholders are of key importance.

However, in some studies, a rights-based approach is seen as the primary mechanism to deliver this, whereas in others, severe top-down controls with very limited participation of fishing communities in the management process are advocated.

We argue here that a necessary condition for successful management contains all these elements: a competent management authority able to set and enforce regulations and monitor the status of the stock, together with some form of rights-based allocation to fishing operators (either collectively or individually) to avoid the situation where overcapacity produces economic hardship and erodes management capacity.

EVIDENCE FROM FISHERIES PERFORMANCE

Reviews of successful fishery management are of necessity specific to individual fisheries and sometimes anecdotal. However, in some large areas, a combination of strong state governance and wealth, substantial scientific activity, and different types of fishery management offer the opportunity of some comparison between different types of fishery management that goes beyond the anecdotal.

Detailed data on the status of different fisheries are published for U.S., Northeast Atlantic, Australian, and New Zealand fisheries. The approach to management taken by these authorities is varied. New Zealand has the most developed and widespread application of individual user rights (ITQs), which have been in place from 1986 and have spawned other developments such as collaborative and alternative research by stakeholders.

ITQs are present also in some Australian fisheries, a very few U.S. fisheries, and some Northeast Atlantic fisheries. (The Faroes, Norway, Iceland, and United Kingdom have rights-based systems, and some fleets, notably the Dutch flatfish and Spanish Grand Sole fleets, are also managed via ITQ.)

If a broad view is taken of these management areas, the evidence for the positive benefits of ITQs in supporting sustainable resource use is mixed. Only 15% of New Zealand's stocks within the quota management system, for which the stock status is known, are substantially below the target reference level.

For other administrations, the percentage of stocks that are below the limit reference level (i.e., overfished), out of the total number of stocks for which the status is currently known, is 19% for Northeast Atlantic fisheries managed by non-European Union (EU) administrations (Iceland, Faroes, and Norway), 25% for federally managed U.S. fisheries, 30% for

Northeast Atlantic fisheries managed primarily by the EU, and 40% for Australian Commonwealth fisheries.

Even within the United States, there are very large regional differences: 40% of major Fish Stock Sustainability Index (FSSI) stocks managed by the New England and Mid-Atlantic Fishery Management Councils, for which the stock status is known, are overfished, and 30% are subject to overfishing. By contrast, only 13% of FSSI stocks managed by the Pacific, West Pacific, and North Pacific Management Councils are overfished, with 6% suffering from overfishing.

Only the United States has seen an improvement in performance over the past several years; in 2000, 38% of U.S. stocks for which the status was known were classified as overfished.

All other areas have experienced some increases in the number of overfished stocks in the past decade, although the increases in New Zealand have been very small and are offset by recoveries in some inshore stocks. However, these statistics disguise a quite dynamic situation within each region; for example, of 74 U.S. stocks requiring rebuilding, biomass is increasing in 48% of them even if they have not yet achieved rebuilt status.

More detailed examination of the U.S. situation reveals that although ITQs are not generally applied, West Coast fisheries are managed by quota controls with fishing rights assigned to fishing companies or sectors, whereas in the northeast, fisheries are managed by a days-at-sea scheme and other effort controls. In terms of their current performance and the stock recovery required by the Magnusson-Stevens Act, West Coast management systems appear to be more effective than northeast coast systems: Only two of the 18 New England stocks that were overfished in 1995 have now recovered, compared to 4 of 9 stocks similarly categorized by the Pacific Fisheries Management Council.

Clearly, non-ITQ management systems do not always fail to maintain sustainable stocks, and management systems using

ITQs are not always successful. The critical additional requirement appears to be a formally adopted management strategy with predefined rules for what to do in different circumstances.

In New Zealand, in addition to an ITQ system, a formal harvest strategy embedded in the Fisheries Act (1996) means that rebuilding is statutorily required when the stock is below its target level.

By contrast, the lack of a formally adopted harvest strategy in the Australian Southern and Eastern Scalefish and Shark fishery has led to an increase in the number of overfished stocks in that fishery over the past 10 years, despite operating with an ITQ system since 1992. EU fisheries also lack a formal harvest strategy; although there is a commitment under the 2002 revision of the Common Fisheries Policy to develop multiannual management plans for all stocks, such plans are only currently defined for 17 of the 94 stocks that fall under EU management, and many of these have had to be negotiated during periods of stock collapse.

The key problems, i.e., the need to provide incentives to fishers to engage constructively in fisheries management and the need to have strong legal support for predefined harvest strategies, apply equally to management of stocks under national control and those in international waters under the control of Regional Fisheries Management Organisations.

To our knowledge, none of the latter currently allocate rights to individual fishers, and only a few have defined and tested effective harvest strategies. Allocation problems continue to beset these high-seas fisheries and influence compliance, data availability, and transparency.

The recovery of depleted fish stocks is a key issue and one to which most countries committed themselves in 2002 as part of the World Summit on Sustainable Development. An effective reduction in fishing effort, the participation of fishers and other stakeholders in the science and decision-making process, and the biology of the species are important factors affecting successful recovery. However, unless a harvest strategy is defined, with pre-agreed, legally binding decision rules requiring reductions of effort when stock sizes decline below limit reference points, most management authorities will still delay taking action to recover stocks.

Some of this delay may arise from uncertainty in the science, but mostly it arises from an unwillingness to take decisions that will create hardship for fishers, and usually a delay will exacerbate stock decline.

Ultimately, the most successful management approaches are likely to combine rights-based systems, creating incentives for fishers to operate efficiently and with long-term sustainability in mind, with a strong legal structure that requires the development of pre-agreed harvest strategies and decision rules that are triggered and adhered to as reference points are passed. As indicated earlier, an adequate control of fishing activities is also necessary.

ADDRESSING ECOSYSTEM-BASED MANAGEMENT

In recent years, there have been many calls for much wider use of Marine Protected Areas to address the need for ecosystem-based management.

We see these as a useful part of fishery regulation, but they are not a universal solution because unless the basic issues of capacity, regulation, and rights are solved, protected areas will simply displace the problem elsewhere. In this we concur with recent reviews that emphasize the primary importance of conventional measures to control fishing mortality and the secondary but essential role that marine protected areas (or other area-based management, such as local prohibitions on particular gears such as bottom trawls) have in dealing with specific issues of ecosystem conservation, such as bycatch and habitat damage.

The simple creation of rights-based incentives does not automatically deal with ecosystem problems, because fishers have little incentive to minimize bycatch or habitat damage that does not affect their target species. An interesting recent development is the creation of additional incentives for fishers through market measures, such as the creation of sustainable fisheries certification schemes and pressure from environmental nongovernmental organizations for responsible fisheries.

Fishers have a major incentive to improve fisheries to satisfy certification conditions, and so far most of the conditions raised in Marine Stewardship Council certifications have concerned the ecosystem effects of fishing, often related to quantifying and reducing deaths of bycatch species and damage to habitat.

Even in the statistics documented for some of those states with appreciable management capacity, what is striking is for how many stocks, the status is uncertain or not determined. In the United States, the stock status of 30% of the 230 major (FSSI) stocks and stock complexes was undetermined in 2006; in Australia (48%), New Zealand (78%), and the Northeast Atlantic (61%), the numbers are even higher.

Given the problems that most authorities have in deriving reliable quantitative assessments of their stocks of major commercial importance, the large numbers of small, commercially unimportant stocks present in most areas, usually as bycatch, cannot realistically be assessed.

Under a comprehensive ecosystem approach, risk assessment methodologies should be used to identify those bycatch species in need of special measures, and monitoring programs, for instance using scientific observers, need to be implemented to monitor trends in all bycatch species. The application of these approaches is in its infancy even in the most advanced management schemes; many simply respond by setting untested but hopefully precautionary effort or catch limits.

These considerations apply even more strongly to fisheries operators in developing countries. In a situation of little or no management capacity, some form of bioeconomic equilibrium is the likely result, but in such cases the management priorities may be different. Indeed, high employment with relatively modest economic rent, as long as it is compatible with the sustainability of the resource, may be a perfectly legitimate management goal. In other cases, the development of Territorial Use Rights (TURFS) within local communities can lead to effective management control and rights-based operations, resulting in successful management.

CONCLUDING REMARKS

There is no doubt that there is a major problem with the world's fisheries, and, despite serious attempts to improve management and to facilitate recovery of depleted stocks, the success has been limited. The key issue that we highlight in this review is that for successful management a dual approach is required, one in which authorities provide incentives for conservation based on fishers' rights and which is supported by strong management incorporating legally enforced and tested harvest strategies.

5

Trends in Fisheries Management

INTRODUCTION

Poorly defined property rights, increasing effort and decreasing catches are resulting in conflict in fisheries throughout the world. While these effects are common and well documented in highly regulated fisheries in the European Union (conflicts over allocation of quota for example) and North America (for example conflicts over access to salmon stocks between the USA and Canada); the effects upon fisheries in developing countries, while under-researched are invariably more dramatic and can have long-term implications for the promotion of sustainable livelihoods. While there is as yet little evidence of a sharp increase in natural resource conflicts (Hussein, 1999) it is apparent, however, that the consequences of conflict in the management of natural resources are becoming more detrimental to long-term sustainable exploitation and are effecting an increasingly large number of people.

Although there is an extensive literature on natural resource conflicts, little work has been done on a) analysing the causes of conflict beyond the case-study arena or b) applying economic tools to the study of conflict. In an attempt to redress the balance, this paper sets out some initial findings from an on-going study into the management of conflict in tropical fisheries. Drawing on New Institutional Economics and common property resource management theory, it analyses how and why fisheries institutions adapt to changing circumstances and the role of conflict in the process. Using evidence from Ghana it examines the emergence of fisheries management institutions under differing access regimes and analyses the factors which appear to have influenced institutional change.

The paper argues that conflict can be the result of rising transaction costs and the inability of natural resource institutions to manage these changes. It also argues that conflict can be both a positive and negative force and should not necessarily be eliminated altogether. The paper is divided into four sections.

First, it locates fishing as an economic activity within developing countries and looks at the interactions between local, national and international policy objectives and how these can impact upon the development and exploitation of the resource. In the second section, conflicts are described and why they emerge in natural resources management is discussed. Focusing on the role of institutions it looks at how changes in institutions might affect the management of common property resources.

Section three draws on initial findings from field research conducted in Ghana between March and May 2000. In the final section, the paper presents a number of conclusions as to the process of conflict formation and the possibilities for managing it. The role of fisheries in developing economies Fisheries play an important role in the local and national economies of many developing countries where they provide food, income and employment for large sections of the population. However, the framework within which they operate is often complex and vulnerable to failure. The absence of clearly defined property rights, the trade-offs between national and local needs and the impact of rising pressures on the economy can, however, all contribute to the failure of fisheries to positively contribute to the development process.

The national perspective: Following the second world war the rapid expansion of fishing capacity in many developing countries was promoted through the so-called Blue Revolution.

New fishing technologies and fishing methods coupled with a desire to improve nutrition and generate foreign exchange led many countries to build on fishing activity already present. Some of this development was as part of national development programmes, in other cases, international loans and funding programmes helped promote the development of national fishing industries. As a direct result of this drive to build up national fishing capacity, the participation of developing countries in global fishing activity rose dramatically between 1955 and the 1980s. The rewards to be gained from participating in the global fishing market were potentially large, and many benefited, albeit in the short-term, from their forays into fishing.

Increased export-revenues and the multiplier effects associated with expanding industries all helped boost the development of many developing coastal states.

The long-term consequences of such developments, have, however, had a dramatic effect. Rising incomes and increased borrowing of cheap petro-dollars contributed to chronic debt problems as the global economy staggered to a halt in the early 1980s; an attitude of positive speculation on fish-stocks, lack of proper assessment and regulation led to the over-exploitation of a number of important stocks (Schurman, 1996; Thorpe et al 2000).

As a consequence of the international debt burden and the strictures placed on the economy by the Structural Adjustment Programmes, trade-offs between local and national needs put further pressure on resources.

The downsizing of the state and the increased role of the market as the resource allocation mechanism saw many developing countries expand their export fisheries to generate revenue whilst at the same time the power to regulate such activity was hampered bythe limited capacity of Fisheries Departments to undertake interventions.

The push for export revenue and the renting of grounds to Distant Water Fleets has resulted in spatial conflicts between industrial and artisanal fleets (Johnson, 1996), state support for local fisheries has been constricted and external economic pressures now drive national policy formation (see for example the work by Stonich, Bort and Ovares, (1997) and Neiland, Soley and Baron (1997) on the expansion of shrimp farming).

From a local perspective: In agricultural-based economies, natural resources form the social safety-net that supports large parts of the population. Access to common property natural resources such as timber, fuelwood, grazing land, fisheries, forest products and irrigation water are fundamental to the livelihoods of many of the world's poor. These natural resources are often held as common property and the survival of rural populations has traditionally relied upon a complex set of institutional arrangements that govern use of and access to these resources.

The formation of institutional access and use arrangements is well-documented (eg.Ostrom, 1994; Alegret, 1996; Feeny, 1988). Whilst many communities have succeeded in maintaining their common property resources through collective action, in some cases pressures internal and external to the community have caused arrangements to collapse (Berkes, 1986; Mensah-Abrampa, 1988; Adam, 1994). The outcome of a resource becoming open-access is likely to be over-exploitation and the dissipation of resource rents (see Cunningham et al, 1995 for example). In terms of development, then, as the ability of a community of users to exercise rights over their resource is eroded so the role of the resource as a safety-net is undermined.

The pressures that lead to the erosion and collapse of common property resources are diverse. They can result from a rapid rise in population (either through natural growth or an influx of migrants); changes to the ecological status quo (through droughts or flooding, for example); national policy changes that affect rural employment rates; technological change that results in higher harvesting potential and social change that can alter income differentials, systems of property rights and market systems.

An important outcome of the breakdown of common property arrangements is the emergence of conflict between stakeholders. However the precise mechanism is not fully understood. This will be explored in the following section. An anatomy of conflict

A review of the conflict literature reveals numerous theoretical approaches to describing and explaining conflict. Conflict is a function of social structure (sociology), of power or class relations (political science) and of individual utility maximisation (economics). It is both positive and negative, constructive and destructive (Powelson, 1972), violent, coercive and non-violent (Wallace, 1993). Thisdiversity of approaches reflects the wide range of disciplines that have addressed the subject. There are many possible definitions of conflict, the following synthesis of definitions is used in this chapter:

Conflict is a situation of non-cooperation that involves groups of people with differing goals and objectives and yet Conflict is dynamic and can be a positive catalyst for change Although the term 'conflict' often has negative values attached to it, this is not necessarily the case.

Conflict can be described as negative when the outcomes are a zero-sum game (no-one benefits) rather than a positive sum-game (everyone benefits) or where there is a deadweight loss of social resources as a result of the 'guns vs butter' argument.

However, it is also important to bear in mind that conflict can be positive and attempts should never be made to eradicate it completely or to prevent it emerging. Conflict encourages goods to be produced more cheaply, government to become more efficient, flaws in the set-up of institutions to be ironed out and allows society to function efficiently by resolving small conflicts often (Powelson, 1972).

At this point, the question of the difference between conflict and competition arises. Competition implies the existence of rules, conflict arises when those rules are, for whatever reason, disbanded or ignored. In order to help clarify whether conflict is positive or negative it is often helpful to look at what the conflict is over and how fundamental that disagreement is to the social status quo.

Aubert (1963), Boulding (1966) and Powelson (1972) all differentiate between conflicts that are 'within consensus' and those that are 'over consensus'. In the former case the parties to the conflict agree about the value of what they seek but not the means of achieving it. In the latter case the parties to the conflict are unable to agree on the value of what they seek nor on how to achieve it.

The potential impact of conflict is thus dictated by the degree of consensual framework within which they are contested and the degree of conflict over basic consensus.

AN ECONOMIC EXPLANATION FOR CONFLICT

Economic theory provides two reasons why conflicts emerge over natural resources:

(a) the allocation of scarce resources requires trade-offs which become increasingly difficult as the demand for and supply of resources changes and

(b) short-term personal gain wins out over long-term social needs or benefits.

The resource allocation issue: Natural resource conflicts occur when the resource in question has become so scarce or degraded as to raise issues of allocation amongst the community of users. Under perfect environmental conditions, as the ratio of users to resource grows so expansion takes place: extra land is brought into cultivation, new areas of forest are exploited, different species are fished or fishery activity moves along the coast or further out to sea.

Powelson (1972:33) argues that so long as the answer to 'who gets how much' is resolved by producing more, conflicts over allocation are positive because they encourage growth.

When the ecological boundaries of the resource have been reached, further expansion is no longer possible (without adverse consequences) and the division of the metaphorical cake has to change because the option of increasing the size of the cake is no longer possible. At this point, resource allocation mechanisms come under increased pressure to satisfy all users.

The equity of resource allocation tends to diminish as the ability to expand production decreases.

As government departments attempt to placate all the stakeholders, so resource allocation decisions have to change and conflict ensues as certain stakeholders gain precedence over others. This is of course nothing new. The trade-offs between equity and efficiency occur in all markets and are faced by all governments. In the developing world however, where democracy may be weak and where powerful interest groups are able to co-opt and unduly influence the decision making process, managing the trade-offs is difficult. Short-term gain vs social needs: There is a large body of literature that uses game theory and behaviour modelling to explain why common property management institutions (in particular) can fail (see for example: Ostrom, 1994; Ostrom et al 1994; or Walton, 1998). The Nash Equilibrium and the Prisoner's Dilemma illustrate this point neatly: in the classic decision making model-the Prisoner's Dilemma-the best possible move for player A given the best possible move for player B is to not co-operate.

Thus the socially optimal outcome of the 'game' is not a Nash Equilibrium (Gibbons, 1992; (Jennings, 1999 pers. comm; Parker and King, 1995; Varian, 1990). McKean (1992:248) notes, in a similar vein that the short term benefits of cooperating are often outweighed by the costs.

Consequently a community opts for the least-cost option which is to withhold their contribution to the collective goal. A third explanation for the emergence of conflict has also arisen from the more recent New Institutional Economics school, this sees institutions and transaction costs as a key element.

From a new institutional economics perspective, institutions exist to minimise transaction costs and are a means of transcending the social welfare dilemmas that arise out of individual action and help maximise collective welfare (Bates, 1998:35). They are also the mechanism for dealing with 'gaps' left by market failure in insurance and risk assessment

(Bates, 1995). Transaction costs are literally the costs involved in negotiating a transaction and have been described as the economic equivalent of friction in the world of physics (Hubbard, 1997). Transaction costs are the costs associated with gaining information, making decisions and carrying out decisions (Abdullah et al, 1998).

In terms of fisheries, they can be divided into the ex-ante costs of collecting information and making collective decisions in the fishery and the ex-post costs associated with implementing collective decisions. Kuperan et al (1999) argue that transaction costs in fisheries arise, mainly, from the fact that the fishery involve multiple stakeholders and differing objectives and long-term goals. Institutions are important as resource allocation mechanisms-markets and fisheries departments, for example, both have a role to play in allocating scarce resources. What is more, it is now recognised that institutions matter for economic development and recent development research has recognised the need to 'get the institutions right' if effective economic development is to be achieved.

Katterman (1998:294) points out that the OECD has finally recognised that technical co-operation, the long-time philosophical foundation of development practice, can amount to nothing if the institutions that it relies upon for success are weak. Thus for both successful natural resource management and economic development institutions able to adapt to changing circumstances are needed. There are two explanations about why institutions might fail in the face of change.

The first relates to the supply and demand for institutional change, the second to transaction costs. Thomson, Feeny and Oakerson (1992: 132) and Feeny (1988) discuss the supply and demand model for institutional change: when the demand for institutional change (to capture gains not possible under existing arrangements) outstrips the ability to supply change, failure emerges. They list relative factor or product prices, the size of the market, technological change and fundamental decisions of government as causes of change that lead to what they term 'institutional disequilibrium'. However, they also recognise that change depends on the State's willingness and ability to help new institutions emerge.

This is a particularly pertinent issue in the developing world where, as discussed before, state capacity may be underdeveloped. Because in the NIE perspective institutions evolve to minimise transaction costs (through minimising uncertainty)-when it is no longer able to do this effectively its position is weakened and it is increasingly unable to deliver services effectively or be allocatively efficient (Klitgaard, 1998:337). Using the NIE paradigm and the supply and demand argument above it could, therefore, be argued that a rise in transaction costs, over and above the institution's ability to accommodate this change leads to inefficient operation. This rise in transaction costs could be due to development pressures (political and economic), environmental scarcity (perceived or otherwise) and structural problems (political and economic) which increase the costs mentioned earlier: the costs of collective decision making, information gathering and collective operation.

In the case of fisheries management, strong and flexible institutions are required at both the national and local level. Arguably, the more nested the structure, the lower the transaction costs between the top and the bottom of the system and the more efficient the institution in its allocation role.

In many developing countries this may not be the case. Communication between the different layers of fisheries management are frustrated, the legitimacy of the ruling body to assign resources is often missing or contested and political factions and the action of rent-seeking elites influence the vertical relationship. Institutions, conflicts and fishing in Ghana The political and economic context

Ghana, the former British colony of the Gold Coast, became the first African colonial state to gained independence in 1957. It has had a number of military coups, but is set, in December 2000 to become the first African state in which a military ruler has handed power to a civilian government through the ballot box. Unlike many other countries in the region, Ghana has experienced comparatively little unrest with its neighbours or amongst its diverse tribal and linguistic groups, and it is perhaps this aura of stability and peace at a national level which influences conditions at a local level. In the 43 years since independence Ghana has made some remarkable advances in development and has also suffered some serious set-backs.

Despite significant advances made in improving some of the standard development indicators (life expectancy has risen by 4 years since 1995; adult literacy has risen by 44% since 1970), Ghana is currently struggling with the dual effects of economic adjustment measures and external shocks. Economic reforms have seen public sector budgets slashed and the impact in the natural resources sector, to name but one, has been devastating. All departments are under funded, and in the fisheries department, this has lead, in particular to a reduced ability to monitor fisheries effectively and enforce regulations. The need to raise revenue has seen the introduction of VAT at 10% which is set to rise to 12.5% later this year. Despite well intentioned publicity campaigns to educate the public as to the need for VAT, this is unlikely to ease the blow of a significant price rise on all VATable goods-particularly on basic grains that have shown sharp price rises since last year. Coupled to this is the rapidly devaluing national currency which, while making Ghanaian exports cheaper, makes servicing the national debt (currently $580 million) more expensive daily and dramatically increases the costs of vital industrial inputs.

Ghana is an oil importer, and recent OPEC price rises have also hit the economy hard. While exports have grown aided by improving terms of trade and export initiatives put in place as a result of economic reforms, these have tended to concentrate on cash crops such as cocoa.

In the context of fisheries, the prevailing economic conditions are having a number of effects. Firstly the day to day costs of fishing are rising. Fuel prices not only impinge on the length of fishing trips, but exchange-rate depreciation has increased the costs of imported inputs such as nets, ropes and outboard motors. The knock-on effects of such price rises are widespread. Education and health at the local level both suffer as more and more income is diverted away from the 'add-on luxuries' into the basic needs for survival. Although things have not reached crisis level as yet, there is evidence of malnutrition in fishing villages and primary schools lack basic teaching materials (supplied by the community). Despite this catalogue of economic problems, some very interesting results emerged from a recent study of the causes and management of conflict in coastal Ghanaian fishing villages.

SURVEY

Between March and May 2000 the Centre for the Economics and Management of Aquatic Resources in collaboration with the Ghanaian Department of Fisheries conducted a survey of 64 fishing villages along the Ghanaian coast. This sample represents 33% of the villages listed in the latest Canoe Frame Survey (1995) conducted by the Department of Fisheries. These villages are spread amongst 4 regions. With no information on the variability of the population it was decided to select one third of all the villages in each region on a random basis. In order to gain a complete picture of conflicts within the coastal zone, interviews were also conducted with inshore-vessel owners and a number of industrial fleet operatives.

Ghanaian artisanal canoe communities are organised around the Chief Fisherman. A complex set of procedures has to be undertaken before any interviewing can take place and no interviewing can be conducted with permission from the Chief Fisherman. Consequently we targeted the Chief Fishermen who on most occasions was accompanied by a number of other village elders, usually comprising his right-hand man and the most skilful or successful fisherman in the village.

On a number of occasions large groups of women traders and other fishermen were present at the meeting. For reasons of simplicity and speed a number of tools were adapted from the extensive PRA 'toolbox' to help build PISCES (Participatory

Institutional Survey and Conflict Evaluation Study) developed specifically for use with artisanal fishing communities to gather information on institutional arrangements.

An Oasis of Calm in a Sea of Troubles

Theory would suggest that given the economic context within which Ghanaian fisheries are operating, the open-access nature of the resource and the inherent instability of pelagic stocks, conflict should be widespread and the fisheries management institutions under considerable pressure. Initial analysis, however, suggests that this is not the case. At the time of writing data collection was still continuing in Central and Western Regions. The summary results below are therefore preliminary results on regions already completed.

Villages as institutions. From the results so far, Ghanaian coastal villages appear to be stable and differ little in their organisational structure along the entire length of the coast. The village chief is at the top of the hierarchy, the Chief Fishermen, first point of call for all fisheries matters sits beneath the Chief.

Although by law the Ghanaian coastal line is open access (to Ghanaian citizens), in practice, a complex set of local laws and a system of reciprocity and responsibility govern inshore waters. 91.7% of villages questioned confirmed that although anyone could fish off the beach, in practice, they must seek permission or announce their intentions to the Chief Fisherman, this includes residents and migrants. The remaining 8.3% stated that anyone could fish off the beach, and did not qualify their answer. No one reported ever having a request to fish denied.

Once permission has been granted by the Chief Fisherman he then becomes responsible for the welfare of the fisherman and his kin. This system of 'responsibility' is not unique to Ghana and acts as a kind of insurance scheme. There are many internal migrant fishermen in Ghana-some are migrants that seasonally move from one area to another, others are migrants that have been settled in a village for generations but, because they are from a different language group, are still classed as migrants.

The distinction between natives and migrants to so powerful than in one village in Western Region, there are two separate landing sites: the Fante-line and the Ewe-line highlighting the two distinct groups that have fished from the village for many years.

The institutionalisation of fisheries management. The local institutional structure of Ghanaian fisheries has, on the face of it, changed very little over the past centuries, and it is perhaps this stability and consistency that helps it remain comparatively peaceful.

However, economic reforms, as described above have caused many significant negative changes to the national organisation of fishing in the past five years but in some areas have also ironically tended to bolster the local structures rather than threaten them. The most significant positive change has been the introduction of the Community Based Fisheries Management Programme (CBFM) under the auspices of the Fisheries Sub-sector Capacity Building Project.

The success of the CBFM would appear to be largely due to the fact that rather than radically changing the status quo, it has strengthened the existing structures.

The Fisheries Sub-sector Capacity Building Project is a joint venture between the Government of Ghana and IDA/ World Bank. Started in October 1995 its main objectives were to improve the long-term sustainability of Ghanaian fisheries. Within this main objective was also the aim to improve the capacity of the Department of Fisheries, address the issue of lack of an active management regime, weak institutional and legal frameworks for fisheries and a growing financial and resource crisis in the industry.

In order for the sustainability of Ghanaian fisheries to be improved management plans were needed and for these to be successful they would need the full approval of local communities. The Project will run until 2001 when the funding will finish and the Government of Ghana will be responsible for seeing its continuation.

As part of the CBFM, committees have been formed in each community and their first task was to draw up a list of by-laws governing fishing activity there. By drawing up their own by-laws and submitting them to the District Assembly for ratification, local norms would be lent weight and legitimacy. This has a two-way function. Firstly the community feels that its laws are valued because they are recognised by a higher authority and secondly, it engenders a sense of trust and liaison between the communities and the District Assemblies.

The make-up of the committees varies but as a rule consists of the chief fisherman and representatives of the various stakeholders. Typically this would include the indigenous and migrant fishermen, the fish processors, fishmongers and fishtraders, vessels, gear and engine owners, and a representative from the Fishermen's Service Centre.

During the course of the field work the villages in a number of districts in Central Region were going through the process of ratifying the by-laws which had recently been gazetted.

The drawing up of by-laws has allowed existing norms to be institutionalised and has allowed 'best-practice' in fishing to be argued out amongst the fishermen and agreed up in a formal setting. A good example of this is the by-law that bans children from the beach during school hours (recognising that schooling is a very important part of village development) and the by-law that formally bans fishing with dynamite (an extremely contentious issue up to this point).

Conflicts and conflict management. In the regions so far covered, violent conflict between fishermen was very rarely mentioned. Day-to-day squabbles and difficulties are, however, common. Interestingly, the pricing of fish is regarded as a conflict, although it could be argued that this is an essential and integral part of how markets work and is a sign of competition not conflict.

However, because the ethos of the survey was for the communities to identify what they considered to be conflicts, we have let it stand. Two other conflicts often attracted heated debate: the incursion of semi-industrial trawlers into water less than 30 metres deep (reported by 25% of villages) and fights among women over access to catches and credit facilities (reported by 29.2% of villages).

Although there was scant reference to a rising number of canoes or fishermen related to the presence of conflict, two examples stand out in particular. One was a fishing community in Accra which suffers from its proximity to the city centre and all the urban poverty problems associated with it. Here the Secretary to the Chief Fisherman mentioned that the landing beach was no longer able to cope with the number of canoes and there were many conflicts over landing canoes and off-loading catches.

The other was Mumford, a well documented fishing village also reported problems with lack of landing facilities exacerbated by the increased number of canoes present in the village. Mumford's case is discussed in more detail later on. Based on the answers recorded, conflict in Ghanaian coastal artisanal fisheries can be divided into a number of categories: conflict that results from outside influences, conflict that results from the internal allocation of resources and that which could probably be better described as competition.

The typology below attempts to categorise the most frequent problems encountered, it should be noted that at the margins some conflicts could happily sit in a number of boxes. Those conflicts caused by outside influences are the hardest to solve because they involve elements beyond the immediate control of the village.

Because violent conflict is almost unheard of, it is perhaps useful to establish a conflict scale for Ghana that better distinguishes between the various degrees of day-to-day squabbles. Conflict management at the village level is highly structured and organised and it is perhaps a reflection of this that conflicts are few and far between. All villages in the survey reported that any conflict between fishermen is reported first to the Chief Fisherman who then, along with his panel of elders, comes to a decision on the issue.

In the case of damage to nets or boats caused by other canoe owners, the culprit is usually made to pay 2/3 of the damages (recognising thus that fishing is a dangerous activity and incidents are rarely deliberate). Although these cases may take a while to resolve-establishing fault and liability-they are not considered tô be onerous. The conflicts that do take longer to resolve are those between different types of vessel-trawlers damaging the nets of canoe fishermen cause a frequent problem.

Even when there are a number of witnesses and broad daylight has enabled the collection of the name and registration details of the offending vessel, the cases can take months to resolve. Such cases are reported to the relevant local fisheries office that then deals with them. Trawler owners often deny all responsibility for the damage and communication and transport difficulties between the administrative centres and the villages further lengthens the process.

During discussions other factors arose that, whilst not identified as conflicts by the villagers, were identified as making life in the villages more difficult. Two examples of these are the incidence of erosion and the rising costs of inputs.

Sea erosion was frequently mentioned in villages in Greater Accra and Volta region to the east of Tema harbour. Sea erosion in the Bight of Benin is serious; current large scale project funded by international agencies are working to building sea defences to protect the coastline. Local lore suggests that without these sea defences Keta Lagoon (a significant water body close to the Togo border) will disappear into the sea in five years time.

The building of Tema harbour was the most frequently cited reason for erosion in the east of the country. The knock-on effects of the Akosombo Dam, built on the Volta River are also cited for disrupting flooding patterns in the river delta (this has resulted in decreasing water levels in lagoons) and the disappearance of certain fish species at the mouth of the river.

The rising price of inputs was universally mentioned as one of the hardest issues to contend with. Given that changing economic circumstances affect how communities deal with the day to day running of their lives, it is hardly surprising that there is a perceived link between increased economic hardship and conflict.

Only increased funding of a monitoring and enforcement capacity and negotiation with the trawler owners is going to be able to better deal with this problem. This is supposed to be covered under the remit of the FSCBP, but as it liable to be a long-term goal. The root cause of the trawler issue-declining catches and rising costs pushing the trawlers into illegal areas-has a number of long-term solutions.

However, the trawler sector is facing severe economic pressures and any measures are probably too painful to contemplate at this juncture and beyond the capacity of the State (economically and politically) to implement currently.

The Community Based Fisheries Management Programme (CBFM) would appear to have had positive benefits to the communities. By using the existing institutional structure, the CBFM has enabled more formalised management to be introduced to communities without upsetting what was clearly a system that worked well to begin with.

The only communities that appeared indifferent to the CBFM were those in urban areas (principally Accra). The reason given for this attitude is that fishermen in these communities tend to be better educated and have greater access to media and other information.

They are more cynical of government promises of how things will get better in the future and more likely to rebuff any attempts by the government to interfere. Outside urban areas, the CBFM was the reason cited for the decrease in conflict or the satisfaction with the conflict management system in place.

The gazetting of local by-laws was looked upon favourably by all those spoken to. So, in terms of transaction cost analysis, it could be argued that providing local institutions with a more 'legitimate' basis both within their own community and in the district as whole has helped maintain transaction costs at a stable level. In those communities where the incidence of conflict has declined, the CBFM may even have helped reduce transaction costs.

There is, however, a corollary to this argument. Although the CBFM has had a number of positive benefits, it has also inserted bureaucratic systems into institutions that worked quite happily without them before. Further research needs to be done on this issue, but it would be interesting to see if CBFM has actually increased the transaction costs of some communities in terms of time spent at CBFM meetings, costs incurred in travel to meetings and the added costs of bureaucracy involved.

CONCLUSIONS

Conflicts arise in all fisheries regardless of size or structure. However, where the community has a degree of control and immediate interest in the health of the fish resource, there is more likely to be a collective will to manage the resource for the benefit of the community as a whole and to thus mitigate the effects of conflicts.

As control of the resource is removed from the community of users, so the incentive to act collectively diminishes and the conflicts erode any sense of 'rules of engagement'. Although much has been written about natural resource conflicts, little is understood about how these conflicts emerge, develop and are managed. In an attempt to improve the understanding of conflicts, this research examines the role of institutions in the process. In particular, it is interested in the contribution that economics can make to the understanding of institutional capacity and conflict formation.

Although physical factors often spark off conflicts: a rise in the number of resource users, the decrease in fish stocks, climate and ecological changes, it is the ability of institutions: both formal and informal to manage this change that is critical. In the case of Ghana, it would appear that strong local institutions, supported, albeit at a minimal level by the state, have helped maintain conflict at a low level, despite the rising transaction costs around them.

A number of issues remain beyond the realm of influence of local institutions-sea erosion and trawler activity in particular. However, the fisheries management mechanism in Ghanaian coastal communities is strong enough to be able to absorb rising transaction costs and internalise them thus keeping conflicts to a minimum.

6

Fisheries and Seafood Industry

INTRODUCTION

Bangladesh was once a fish surplus country about half a century back when the population was less than 20 m. But with the rapid population growth fishing efforts increased greatly while culture and conservation were very insufficient to keep pace with population explosion.

In the meantime due to high demand in the world market for shrimps, prawns and other crustaceans, Bangladesh stepped into a new era of sophisticated industrial processing development. The evolution of the shrimp processing industry in Bangladesh dates back to 1959 when the first fish and shrimp processing and freezing plant was installed in Chittagong by 1971, there were nine such processing plants in the country with a total production capacity of 58.5 MT per day. All these plants were mostly engaged in processing and exporting of fresh water headless shell on shrimps to Europe and U.S.A. During the Bangladesh War of Liberation in 1971, there was a short pause in the development of the sector. From 1977 onwards the trend of development of the industry started regaining its momentum and by the end of 1992, there were ninety seven processing plants in the country with a daily freezing capacity of about 680 tons. All these plants were primarily designed for processing of shrimps and froglegs. The processing of fin-fishes were not taken into consideration at the beginning in view of the national requirements. But due to the rapid development of the industry, raw material constraints in shrimp and froglegs were felt very much, thus forcing many of the plants to diversify products in favour of fin-fishes also.

A new technology for the production of salted and dehydrated fish specially for sea-water jew fish was also introduced in collaboration with Hongkong and Singapore buyers which helped the country earn a good amount of foreign exchange. The developments in the fish and shrimp processing sector helped the country's export to boost up from a meagre US$ 3.06 m in 1972-73 to 131 m in 1992-93 an increase of 4281% in 20 years. Export of shrimps and fishery products stands next to garments, jute and jute goods and leather. It contributed to 6.50% of the total export earnings of US$ 1994.00 m in 1991-92.Inspite of all these developments the export-oriented sea foods processing industry of Bangladesh has been passing through a severe crisis due to acute shortage of raw materials. The industry has grown at a much faster rate than that of the growth of the raw material procurement. To overcome this situation, topmost importance and priority have been given for the development of fisheries resources specially of the aquaculture resources of fish, shrimps and other crustaceans for which Bangladesh has a great potential.

PRESENT STATUS

Present status in respect to the growth of the Fish and Shrimp Processing Industry vis-a-vis the availability of raw materials which are the living aquatic resources such as freshwater fish, marine fish, frogs, prawns, shrimps and other crustaceans, deserves a brief explanation in this context.

Fishery Resources

The Bangladesh Fisheries can broadly be divided into (i) Inland or Freshwater Fisheries and (ii) Marine Fisheries.

Inland Fisheries : The inland fisheries of Bangladesh are considered to be extremely formidable in terms of natural water areas and its potential for shrimp and fish culture. Inland fisheries contributed to 74.90% of the total fish catch of the country in 1992-93 and can be defined into the following categories.

Open Inland Water Fisheries: It includes innumerable rivers and their tributaries, baors, haors and the estuaries. The main river systems in Bangladesh include the Meghna, the Padma, the Jamuna, the Brahmaputra and the Karnaphuli and their tributaries-the total water area of which is over, 1.03 million ha. There are many ox-bow lakes locally called 'Baors' formed due to silting up of old rivers in the districts of Jessore, Jenaidah, Court Chandpur, Kustia and Faridpur with a total water area of about 5,488 ha.

The natural depressions of land are used partially as agricultural lands in dry seasons and seasonally or perennially filled with water from adjacent rivers during rainy season. Most of these haors are located in the greater Sylhet, Mymensingh and Faridpur districts. Some of the haors are very big. Hakaluki haor and Tangua haor in the greater Sylhet district have water areas of about 36,437 and 25,506 ha. respectively.

The total water areas of the haors in Bangladesh are about 1,14,161 ha. The Kaptai Lake one of the largest man-made lakes, consists of 68,800 ha. Besides there are about 2.83 million ha. of seasonal floodplain areas. The total open water areas of the country is about 4.05 m ha. from which a total of 533,000 MT of fish was caught in 1992-93. This is a very low production which may be increased substantially with proper policy planning, serious efforts and implementation of strict conservation and management methods.

Closed Water Fisheries : It includes large-sized ponds called dighis, ponds and tanks. In most of the areas of the country, almost every homestead has one or more ponds or tanks used for bathing and cleaning as well as for fish culture. There are about 1.29 million ponds covering over 1,46,890 ha. of water areas of which about 76,632 ha are under fish culture, 48,814 ha culturable and 25,450 ha. of derelict ponds. Of the total water areas of the ponds 52.17% are now under fish culture, 30.51% easily culturable but now idle and 17.32% are derelict which can be turned into good fish ponds after proper renovation them. Out of the total number of 1.29 million ponds about 46.48% are under culture, 29.90% culturable and rest 23.62% are derelict ponds. The estimated fish production of all these ponds were only 16,6100 MT in 1986-87 and 24,2572 MT in 1992-93. Even freshwater Shrimp (M. Rosenbergii) can be cultured extensively to meet the demand of the processing industry. Total production of fish from freshwater fisheries were 607645 MT in 1988-89 which increased to 775472 MT in 1992-93. This may be doubled or tripled with concerted efforts of the public and private sectors.

Marine Fisheries: Bangladesh has a coastline of 480 km along the North and North-East part of the Bay of Bengal. It has an internal estuarine water area of 7,325 sq. nautical miles upto 10 fathom depth baseline, territorial waters of 2,640 sq. nautical miles from the baseline, EEZ of 41,040 sq. nautical miles and the continental shelf of 2,480 sq. nautical miles. The total of marine water areas is about 48,365 sq. na. miles which is almost as big as the country itself.

All these water areas have great potential for seawater fishes and shrimps. The coastal area has a great potential for seawater black tiger, white and brown shrimps. Based on different surveys and research works in the Bay of Bengal, it is estimated that the standing stock of fish is around 2,64,000 to 3,73,000 MT and that of shrimps around 9,000 MT. The marine catch increased from 95,000 MT in 1975–76 to 250480 MT in 1992-93 an increase of about 265%.

This has been possible due to Government's encouragement for the introduction of a deep sea fleet of 70 trawlers (out of which 56 are in operation now) and over 6,000 mechanized boats in the Bay of Bengal. Yet, there remains much unexplored areas for development of off-shore pelagic fishing.

Bangladesh water-bodies were surveyed by the Bangladesh Space Research and Remote Sensing Organisation (SPARSO) by using aerial photographs and satellite imagery under the Fisheries Resource Survey System Project of DOF and FAO/ UNDP DURING 1981–85.

The Shrimp Fisheries: In the coastal areas of Satkhira and Khulna districts. People used to make dikes or embankments along the banks of estuarine rivers and allow sea waters carrying shrimp fry or juveniles to enter into it wherein shrimps used to grow under natural conditions without any supplementary feed.

As a result production output had always been very poor. Shrimp production in the area rotates with paddy cultivation in a systematic manner. During pre-liberation period (1971) there were only 2500 ha. of land under cultivation in the Khulna/Satkhira Region with a production of 20–50 kgs per ha. The Government of Bangladesh have taken up many schemes for the modernization of the shrimp culture in the country from the mid 1980's as.

The land under coastal shrimp culture increased from 51,812 ha in 1983–84 to 108280 ha in 1988–89 which remained almost static till 1992–93. Production of shrimp during the same period also increased from 8386 MT in 1983–84 to 18,325 MT in 1988–89 and to 26000 MT in 1992–93. Average production/ha was 85 kg 160 kg and 240 kg respectively during the same period. It appears that the average size of ponds is very big and not upto the international standard. Such ponds offer bad management. per ha. production is also too low as compared to may Asian countries. Farmers are operating their farms on extensive culture methods and do not or very little apply artificial feed and modern methods of shrimp culture.

Modern shrimp culture methods are being practiced by a handful of industrial farmers who have been able to produce upto 3000 kg/ha. annually. Govt of Bangladesh is stepping up measures to increase per ha. production of shrimp to 1000 kg, in which case the total culture based production of sea water shrimps may exceed 100000 MT within a few years.

With the introduction of coastal shrimp culture and deep sea fishing trawlers in the country shrimp catches has been steadily increasing during the last years. In 1983–84 the total shrimp production was 57656 MT which increased to 70,872 MT in 1986-87 and 99,458 MT in 1992–93.

Out of trawler catch of shrimps in 1987–88, species-wise composition was 17.00% of Black Tiger, 8.00% of White, 53.00% of Brown & 22.00% were mixed varieties of smaller sizes. Catch Composition of coastal aquaculture is mostly Black Tigers. Recent information indicate that China, Indonesia, Thailand, Malaysia, Philippines have been producing over 1200–1500 kg/ha. All these developments were possible due to technological improvements for production of shrimp fry through hatcheries and feed manufacturing.

Bangladesh is far behind in artificial propagation of shrimps, as a result of which it is still dependent on collection of wild fry from the coastal estuaries. The Govt. of Bangladesh have taken up few projects for establishment of hatcheries for P. Monodon and M. Rosenbergii. These hatcheries have very recently gone into test-production.

One private hatchery is already in production in Cox's Bazar. Projects such as establishment of demonstration farms; expansion of improved culture techniques, upgrading of the existing shrimp farms in Cox's Bazar and 1430 ha in Khulna region etc. have already been approved by the Govt. and are now at the final stage of implementation.

All these projects, if implemented properly, may help shrimp production improve substantially. But one thing is almost certain that by expansion of the shrimp culture horizontally, the country will derive no tangible result unless availability of shrimp seeds and feed together with other facilities are made available to the farmers for an urgent switch over from the traditional extensive towards semi-intensive or intensive shrimp culture.

Potential for Culture of Shrimps

It is very clear that Bangladesh has a great potential for shrimp culture. It's coastal lands offer a readily available over a hundred thousand hectares of shrimp ponds where P. Monodon and other marine species can be cultured. There are also over one hundred and twenty thousand hectares of fresh water culturable ponds where *M. rosenbergii* and other freshwater species can easily be cultured.

Potential for Coastal Cultured Shrimps: According to recent surveys, there are about 108280 ha. of brackishwater coastal lands under shrimp culture for Black Tiger, white and brown shrimps. Present average production of about 240 kg/ ha being too low, there are enough scopes to improve this a fewfold. According to my estimates the potential for coastal shrimp culture s to the minimum of about 75,000 MT and a maximum of 1,50.000 MT worth 600 m to 1200 m US dollars by the year 2000.

Potential for Freshwater Cultured Prawn

Bangladesh inland open waters have been producing best quality freshwater giant prawn locally known as "Golda" (*M. rosenbergii*) and other freshwater species from rivers, canals and paddy fields as wild catch only. Artificial propagation of the freshwater prawn is still being experimented with limited success. In the meantime, enthusiastic farmers have gone for monoculture of the giant prawn in their backyard ponds by collecting wild fry and have reported encouraging results. There are about 1.29 million culturable ponds.

These ponds have an water area of 120000 ha, where giant prawns may be cultured without remodelling many of the ponds. A 10 year (1991–2000) projection for utilizing these water areas and raising prawn culture production was prepared by me. It may be seen from the projections that in addition to the present wild catch of approx 50000 MT of freshwater prawns, the country is capable of producing 1,25,000 MT of cultured freshwater prawns worth 875 m US dollars by 2000 at 1990 price.

Total Potential of Cultured Shrimp/Prawns: It is evident that Bangladesh has the potential of increasing its shrimp and prawn production, setting aside other crustaceans, to 67,500–1, 05,000 MT by 1995 and 2,000,00–2,75,000 MT by 2000, worth of US$ 500 m to 800 m and 1400–2100 million dollars during the same period. Even if 50% of this projection can be attained, the country will be able to produce 50,000 MT worth 400 m and 1,00,000 MT worth 800m US dollars by 1995 and 2000 respectively.

The Status of the Sea Foods Processing Industry

Number of Fish and Shrimp Processing Plants : Bangladesh has developed a very impressive sea food processing and freezing industry over the last 10–15 years. There were only 9 processing plants in the country with a total freezing capacity of 58 MT daily in 1971.

From 1972 to 1976 only 4 plants with a combined capacity of 44 MT were commissioned. The trend of installation of freezing plants speeded up since 1977 and reached its climax during 1986–1989 period when 39 plants were commissioned in 3 years time Year-wise installation and commissioning of the freezing plants in Bangladesh with capacity.

Capacity Utilisation (Shore based Plants)

The utilisation of the rated capacity of all plants has always been unsatisfactory. During 1960s most of the plants used to run 40–50% capacity. Gradually, the capacity utilisation has decreased due to unplanned growth of the industry and scarcity of raw materials.

Percentage of capacity utilisation. It has been assumed that processing plants in Bangladesh may operate for a maximum of 200 days a year on the basis of shrimp seasons. It is observed from the data given here that the utilisation of plants capacity have decreased gradually from 24.10% in 1985–86 to 16.68% in 1989–90 and to 19.00% in 1992–93.

Overgrowth of Processing Plants (Shore-based)

From the records the existing exportable raw materials which may be estimated at 21–25 thousand metric tons of shrimps, and fish, one thing becomes very clear that only 15–20 units of processing plants can handle the product and the export business at 80–100% capacity, 30–40 units at 60% capacity and the rest of the plants are overgrowth and have grown in a very unplanned and haphazard manner.

The industry has grown at 500% higher capacity as compared to the available raw materials and as such has become a very sick industry. According to export statistics, only 32 units were in production during FY 1992–93 and the rest were simply out of production, only 16 of 32 units earned over 100 m Taka in 1992–93.

Reasons for Overgrowth

The main reasons for overgrowth of the industry are

(i) unplanned sanctioning of loan without any feasibility study

(ii) keeping the industry in the free-list for a long period of time

(iii) liberal financial support from the financial institutions

(iv) hurried and haphazard approach for earning foreign exchange by many

(v) to get bank loans easily and quickly.

Development of Trawler-based Processing Plants (Shrimp Trawlers)

Modern shrimp trawling was first introduced in 1979 by using deep sea freezer shrimp trawler. Before that Govt. trawlers of the Bangladesh Fisheries Development Corporation used to catch limited quantity of Shrimp by employing its refrigerated (Iced) trawlers. By 1992 there were as many as 48 shrimp trawlers in operation almost all in the private sector except 2 in the public sector.

Raw Materials Situation

The Shore-based plants are dependent mostly on brackish water cultured shrimps or wild freshwater prawns such as *M. rosenbergii*.

There are 20000 MT of headless exportable shrimps available in the country, though the demand for raw materials (shrimps mainly) is 1,36,000 MT. The present supply of shrimp is only 15% of the plants' requirement. Shrimp potential out of the existing brackish water shrimp culture farms of 108,280 ha, is about 1,00,000–1,50,000 MT at a modest semi-intensive rate 1000–1500 kg/ha.

Major Problems for Shrimps and Fish Culture

Major problems retarding the growth of fish industry by culture of shrimps and fish in farms, ponds and tanks are as follows:

i. Acute shortage of shrimp feed and fish feed. The country needs over 100,000 MT of shrimp & fish feed, against which only 6000 MT are available locally. More Shrimp Feed Mills are urgently needed.

ii. Acute shortage of shrimp fry. More and more modern shrimps hatcheries and nurseries are needed.

iii. Lack of modern technology.

iv. Lack of funds/loans from financial institutions.

Waste Utilization

At present about 25000 MT of trash fishes are thrown overboard by the fleet of 67 trawlers, thus destroying fishing grounds and polluting the environment. There is no system of waste utilization from the processing plants, fish markets, poultry farms, cattle farms and slaughtering houses. A few modern waste processing plants are needed for making feed for the fish culture projects, especially for African or Thai catfishes which are going to make a major break through in near future.

Contribution to GDP

Contribution of fisheries sector to GDP is around 3.0%.

Employment

The whole of the fisheries sector employs about 1.20 m people who are directly or indirectly dependent on fishing, fish farming, fish processing etc. This figure is about 10% of the total population of the country.

Out of the total about 6,95,000 are inland fishermen for whole time fishing or subsistence fishing, 4,12,000 are marine fishermen, about 5000–6000 are in the processing industry (both regular and casual) and about 87,000 in shrimp farming, fry collection, crab and frog collection, dry and dehydrated fish industry and in fish carrier boats.

REVIEW OF GOVT. POLICY TOWARDS DEVELOPMENT OF FISHERY SECTOR

There has been no Govt. policy for development of the fisheries sector so long. As a result the sector suffered much in respect to a balanced development. For instance, the fish and shrimp culture sub-sector is still in an infant stage, though there is a great potential for improvement. On the other hand, the processing industry witnessed such a faster growth that today about half of the total number of processing plants are out of production and the rest are under-utilised.

Recently Govt. of Bangladesh drafted a "Fishery Development Policy" for coordinated development of the sector. The salient features of the policy are as follows:

Objectives of Fishery Development Policy

- To reduce the gap between supply and demand of animal protein by increasing fish production.
- To create additional employment in fish and fish related industries for the improvement of socio-economic conditions of fishermen-community in rural areas.
- To earn more foreign exchange by increasing export of fish and fish by-products.
- To develop public health and environmental conditions.

Legal Status of Fishery Development Policy

Govt. semi-Govt. Multinational Organisations, Private, Voluntary Organisation, Individual or Groups of Individuals those situated within the geographical area of Bangladesh and are related to fishery development, export, import or fish business must abide by the Fishery Development Policy.

Jurisdiction of Fishery Development Policy

- General Policy.
- Inland Fish Culture Development Policy.
- Inland Open Water Fish Production Policy.
- Sea Fish Development Policy.

- Shrimp Culture Policy.
- Research Policy.
- Training Policy.
- Organisation Policy of Fishery Sector.
- Loan Policy.
- Miscellaneous Policy.

Under the new fishery development policy, private sectors are given priority to develop shrimp culture and fish processing industries.

Foreign investors are allowed to undertake joint venture with the Bangladesh counterparts. At present, no joint venture projects in the field of trawling, shrimp culture are in operation. Foreign investment in shrimp hatchery, nursery and food mills except the processing industry may be of immense help to the country.

Encouragement to Private Sector

The sectoral infrastructural development had been very rapid during the last decade, specially during 1985–90 period. This has been possible due to the Govt. policy for declaring this sector under free list.

As a result, to-day there is a strong infrastructure of (i) 48 deep sea shrimp trawlers, (ii) 19 fin-fish trawlers, (iii) 97 processing plants, (iv) over 6000 mechanised fishing boats and (v) 108000 ha of water areas under coastal shrimp culture etc.

- The Govt. has declared Fisheries as an industry. The Govt. financial institutions and commercial banks provided funds or loan for processing units. For encouraging the private sector, Govt. of Bangladesh has taken the following measures.
- Bank loan at reduced rate of interest for purchase of raw materials.
- Included frozen foods in thrust sector.
- Liberal allotment of foreign exchange for exporters for foreign travels for market study.

- Introduction of credit cards to exporters for foreign travels.

Processing plants established during recent years are of international standard. But some times lack of proper knowledge and negligence towards plant sanitation and personal hygiene of the workers gave rise to quality problems. This problem can be resolved by proper supervision and vigilance. The group training of supervisors and workers may yield positive results.

Solar dried salted and dehydrated jewfish is now being produced mainly in Cox's Bazar for exporting to Hong Kong and Singapore. The dried Jewfish is produced under the direct supervision of Chinese technicians and as a result no quality problem is normally encountered. Dry fish and fermented fish are produced for local consumption by traditional methods. Fish meal and fish products are produced mainly by BFDC for local consumption.

Fish meal is produced from trash fish from trawler catch and from small mixed dry fish from off-shore islands. Fish products such as burger, finger, cutlet, cake, balls and minced fish are produced in a limited scale from low-cost under-utilised fishes by BFDC trawlers and have become popular to the city dwellers.

In order to go ahead with the latest developments being made in the field of fish processing and fish feed manufacture in the highly developed countries, diversification of products by the industry in respect to the following fish foods have been identified for immediate attention. The technology transfer in these fields may be useful to Bangladesh.

Cooked & peeled shrimps, cook-freeze fish products, fish pastes and spreads, marinated fish products, fish protein concentrate (FPC), fish sausage from minced fish, ready to eat cooked fish food packed in ovenable pouch or tray, canned fish in edible oil, brine or tomato sauce, smoked fish, chilled fish, fillet and fish steaks under modified atmosphere packing (MAP which means replacement of air by a mixture of CO and N in the plastic packets), processed shark fins, meat and liver oil etc.

The industrial uses of fish are also of great importance. The possible fields are listed for consideration and development through modern technology transfer from developed nations: leather from shark skins, fish liver and body oil for industrial and pharmaceutical uses, fish silage, animal and pet foods, ornamental and decoration items from fish skin, shells scales, bones, teeth etc., pearl essence from scales of mainly pelagic fish, gelatin and isinglass from fish air-bladder, liquid fish glue from fish skins and heads, shrimp feed etc.

SKILLED MANPOWER

At present Bangladesh has enough trained manpower for operation, repair and maintenance of fish freezing plants, ice plants, cold storage, net factory, fishing trawlers and mechanised fishing boats. All refrigeration plants in Bangladesh have been and are being established by local consultants, engineers and contractors.

Marine Fisheries Academy in Chittagong provides professional training facilities to fishing trawler personnel. Many of them also work in shore-based processing plants. 619 cadet officers in the following fields of fisheries sector have passed out from 1973–74 to 1991–92.

Besides, BFDC also trained many skilled fishermen/sailors/ mechanics for manning the trawlers, mechanised boats and ice-plants and freezing plants. Govt-run Polytechnic Institutes produce diploma engineers and technicians for operation and maintenance of processing and refrigeration plants.

Both public and private-owned vocational training centres produce refrigeration operators and mechanics in huge numbers. Fisherman's Cooperative Society Provides training facilities to the fishermen of mechanised boats. Marine Science Institute of Chittagong University produces marine biologists. Marine biologists are capable of undertaking modern fish/shrimp culture as well as various research works.

Mymensingh Agricultural University provides higher studies in Fish Aquaculture, Fish processing and Handling technology. But practical short-term training for management and operation of different types of fish food and by-products processing plants may-be needed in future in terms of foreign technical assistance.

Field of Study		No of Cadets Passed-Out
Marine Engineering	..	247
Navigation	..	176
Fish Processing Technology	..	87
Gear Technology	..	76
Refrigeration Engineering	..	24
Radio Electronics	..	17
Electronic Engineering	..	16
Boat Swain	..	05
Total		619

PACKAGING INDUSTRY

The packaging Industry in Bangladesh mainly produces simple types of cardboard for master cartons and duplex papers for minicartons. Materials for the packaging is not of international standard. Foreign collaboration in this field will be of immense help.

7

Value of Aquatic Products in Ecosystem

Fisheries sector forms the bread and butter and nutritional source for millions of Indians. Post harvest loss of resources is an area of major concern in this sector and one of the solutions for that is production of value added products; which also can contribute to the improvement of the economic status of the poor.

To address this issue, Technology Information, Forecasting & Assessment Council (TIFAC) has initiated various projects on value addition with the involvement of women Self Help Groups (SHG's) with the technical and logistical support of Fisheries Institutes as well as State Fisheries Departments.

Due to the novelty of the technology adopted, the value added products resulted from the projects were of superior quality and had greater market acceptance. The success of these projects seeks the replication of similar approach throughout India.

BACKGROUND

Consumption of fish and fishery products formed an important dietary practice of the Indians from time immemorial. It has significantly contributed towards the improvement of the nutritional status of the populations. Aquatic resources are vital sources of nutrients, vitamins and minerals and its harvest, handling, processing and marketing provide livelihood for millions of people as well as providing valuable foreign exchange earnings to our country.

Indian Fisheries Scenario

India is blessed with a coast line of 8118 km with an Exclusive Economic Zone (EEZ) of 2.02 million square km and a continental shelf of 0.506 million sq. km.

The inland water resources includes 191024 km of rivers and canals; 2.05 million ha of reservoirs; 2.254 million ha of ponds and tanks; 1.3 million ha of oxbow lakes and derelict waters; and 1.24 million ha brackish waters.

India has an estimated fish production potential of 8.4 million tons, of which the marine sector forms 3.9 million tons and the inland sector 4.5 million tons. As against this, the current fish production is 5.96 million tons *i.e.,* 2.89 million tons from marine sector and 3.07 million tons from inland sector (Vannuccini, 2003).

Fisheries provide employment to about 5.96 million part time and full time fishermen in India. The sector also provides sizable employment to the people in ancillary industries like boat building, gear design and fabrication, fish processing, marketing etc.

The contribution of fisheries sector to Gross Domestic Production (GDP) is 1.5% and 5.0% to agricultural GDP. The per capita fish availability in India is 4.7 kg/ year (Laurenti, 2002). Processed food exports were at over Rs. 13,500 crores in 1998-99; out of which marine products accounted for over 34% (APEDA, 2000).

NUTRITIONAL SECURITY- ROLE OF FISHERIES SECTOR

Fisheries sector can play a vital role, as a potential source, in attaining nutritional security in India.

The current production of fish in India forms only 71.0% of the total potential and hence there is an ample scope of improvement and thus can be added up to the nutritional security.

It is estimated that globally, about a third of the fish catch is not utilized for human food consumption because of post harvest loss resulting from poor handling and preservation.

In India, there is also the problem of under utilization of by- catch due to the many species that are caught in the net.

The cost effective and efficient utilization of aquatic products demands proper processing and distribution.

Demand for fish and fish related products are increasing day by day in our country and reduction in post harvest losses can make a major contribution to satisfying this demand, improving quality and quantity for consumers and increasing income for the producers.

Thus, improvement in the post harvest utilization of fish catches can ensure further nutritional security among a wide range of people in our country.

Viewing fish primarily as a source of protein has been the dominant perspective in studies on nutritional security. This view is primarily not in the right direction, since it fails to highlight the critical role of fish as a wholesome source of Poly Unsaturated Fatty Acids (PUFA's); minerals such as calcium, phosphorus, iron; vitamins like A, B_1, B_2, B_{12}, D etc.; and trace elements such as iodine and zinc. These attributes makes fish a vital contributor to nutritional security of the most deprived and vulnerable populations.

For them, easily digestible fish is indispensable; other natural animal or vegetable protein sources are poor substitutes on both nutritional as well as economic grounds.

ROLE OF TIFAC IN FISHERIES SECTOR

Technology Information, Forecasting & Assessment Council (TIFAC) is an autonomous society under the Department of Science & Technology (DST), Govt. of India. Following the Technology Policy Statement of 1983 and the Technology Policy Implementation Committee recommendations, TIFAC was set up in 1988 for keeping technology watch on priority areas and to promote actions.

A landmark achievement of TIFAC has been the first major national long-term technology forecasting exercise known as Technology Vision for India up to 2020. Major thrust of TIFAC activities includes undertaking selected projects in identified sectors among which Agriculture and Agro Food Processing forms a major component.

Under the Agro Food Processing sector, the fisheries sub sector was constituted under the chairmanship of Dr. S.A.H. Abidi, Member, Agricultural Scientists Recruitment Board (ASRB), Indian Council of Agriculture (ICAR), New Delhi, and few other national fisheries experts as members, with a view of transforming the cottage level fisheries industry, by:

(a) establishing commercial level plants, with high replication potential, for processing/ producing quality fish products to capture enhanced domestic and foreign market and

(b) creating job opportunities for local community utilizing the local resources through science and technology inputs.

This article reviews the success stories of TIFAC fisheries projects on development of value added fish products as well its future targets.

The Projects

To ensure rapid socio-economic development, application of science and technology should be the foremost priority. In this background, TIFAC has initiated various projects on the development of value added aquatic products in different geographical areas of our country with different women SHG's/ Cooperatives in collaboration with fisheries institutes as well as state fisheries departments; who are providing the technical and logistical support to the projects.

The projects are selected and approved by the Fishery Panel based on the technical as well as economic feasibility of the projects, as recommended by the experts in the concerned areas. A Project Review and Monitoring Committee (PRMC), consisting of experts nominated by TIFAC, are constituted for each project for reviewing and monitoring the progress made by each project. The PRMC will review the progress relating to the project and provides technical advise from time to time at various stages of completion. The PRMC after closely monitoring the developments in the projects may provide mid course remedial actions, if required, to overcome the unforeseen obstacles.

OBJECTIVES

The objectives of the projects are to establish a commercial level cottage industry by organizing women fishers and to improve upon their socio-economic status; to improve the utilization of low value fish/ shellfish by value added product development; improvement in quality & quality testing of the product; and market linkage.

Technologies Transferred

The projects have facilitated and efficiently transferred the technologies such as fish processing, preservation, packing, quality control and marketing as developed by the Fish Processing Laboratory of CIFE. This was done through training programs. The training and exposure imparted during the programs enabled the society to address various product quality related issues through the implementation of quality assurance techniques. The advantages of the technology adopted include superior quality, greater shelf life, and higher acceptability. Various value added fish products were prepared from low cost fishes. The products include:

1. *Prawn Pickles*: Pickling is one of the safest means of easy preservation of fish/ shell fish. Pickles are good appetizers and they add palatability to the starch based blend to taste Asian dishes, besides being highly nutritious. The technology involved is simple and can be imbibed by rural women with great ease. A new recipe has been developed, by CIFE, to improve the acceptability and the shelf life. Novel methods of preparations were used to get pickle with good eating quality.

2. *Fish Papad*: Papad is a direct product commonly preferred all over India as a side dish. Fish is incorporated in papad to make papad more nutritious and tastier. The flour of black gram pulse (Urad dal) is used as the major ingredient. Fish was further added to make papad more nutritious and tastier.
3. *Fish Chakli*: Piston or ram type extruder is used in different regions of India to prepare starch or pulse based fried snacks. Chakli (spiral) is one such popular product in Maharashtra. Fish meat is incorporated in chakli to enhance its taste, flavor and nutritive value.
4. *Fish Sandwich*: Using minced fish meat a product was prepared, which could be spread between bread pieces to make fish sandwich.
5. *Fish Paste*: The product is prepared by using fish flesh (minced) with spices and preservatives; which can be used as a side dish along with rice or bread having a shelf life of three months.
6. *Fish Noodles*: Fish flesh is made in the form of paste and then mixed with refined flour to produce delicious fish noodles.
7. *Fish Puff*: By using twin screw extruder a directly eatable cooked product containing ingredients like fish flesh, rice and gram powder with salt and spices can be prepared with a shelf life of six months.

Project Achievements

The projects have successfully demonstrated the potential for processing low value fish efficiently and incorporating them in acceptable, nutritious and safe, novel and traditional products in a small scale commercial cottage industry. The projects enabled the poor fisherwoman to organize and to form SHG's for attaining sustainable livelihood. Training programs were useful in respect to give SHG's an insight into modalities of organizing product processing and marketing. The project helped to create awareness on the potential for value-addition and market diversification among a wide segment of the local people, thus facilitating the technology to get replicated in other parts of the country.

The fishery resources have been utilized maximum leading to reduction of post harvest losses. Due to the higher value of value added products, when compared to low cost fish has earned the women group a handsome profit thus improving their socio-economic status. Stringent quality checking by CIFE resulted in production of value added products with superior quality, thus enhancing the shelf life and profit. The projects facilitated closer interaction between local traders through visits to major markets and trade fair participation. Continued buyer-seller matching, and provision of market information assisted in market development. At the end of the project period, the societies have shown an improved access to major markets through expanding their value added product range.

Other Targeted Areas

The success of projects in the area of value added fish products from low cost fishes have prompted TIFAC to replicate it in other areas and also to initiate projects in other areas also.

The major thrust areas which TIFAC is focusing for its future activities include breeding and culture of ornamental fishes, production of sheedal; and caviar and fillet production from trouts.

1. *Ornamental fish breeding and culture*: Ornamental fish keeping and its propagation has become an interesting activity for many, providing not only aesthetic pleasure but also financial openings. It has reached new heights in the recent past and has gained accelerated momentum, thus providing employment opportunities to many including unemployed youth, women and other unskilled masses. Indian waters possess a rich diversity of ornamental fishes with over 100 varieties of indigenous species, in addition to a similar number of exotic species that are bred in captivity.

   The strategies to be adopted for boosting commercial production of ornamental fishes includes technical support and dissemination of recent research developments in breeding and culture; formulation and preparation of nutritious feed; health management; water quality management; and training program for creating scientifically and technically skilled man power

2. *Sheedal Production*: Sheedal is a form of traditionally fermented fish product prepared in the North-eastern region of India. Apart from its delicacy and food value, people of this region like sheedal very much because of its medicinal value for preventing stomach related disorders and malaria. The process of preparation varies from place to place and community to community. *Puntius spp.* is the major species of fish used for sheedal preparation.

   For sustainable production and uninterrupted marketing of sheedal the approach involves training, mass production, quality up-gradation, and market development strategies.

3. *Caviar and Fillet production from trout*: Caviar and fillets are premium products which have high demand and fetches greater value in the international market. Caviar is produced from the roe of trout and filleting provides quality trout flesh. Trouts are fresh water fishes seen in cooler climate and high altitudes. Jammu & Kashmir is blessed with trout resource and has been identified as the ideal place to launch such a project.

For initiating a project on caviar and fillet production, the major identified inputs includes, sufficient raw material, sophisticated processing equipments, trained man power to operate the project, and recent technologies and research developments.

Aquatic resource is a vital tool for transforming India into a nutritionally secured nation. Development of value added aquatic products can bring both improvement of the socioeconomic status as well nutritional security in our country. The efforts of TIFAC in this direction have shown fruitful from the encouraging results obtaining from various projects. By producing quality value added products and establishing the market, the SHG,s can help India to march towards prosperity as well as self sufficiency.

INTEGRATED FISHERIES PROJECT (IFP)

Established as Indo Norwegian Project during 1952 this was renamed as Integrated Fisheries Project (IFP) in 1972. The Project is engaged in the development of technologies for harvesting and post-harvesting of marine fish resources. It has a well established Fishery complex consisting of a fishing fleet, a modern mechanical workshop and Slipway to slip vessels up to a displacement weight of 250 tonnes, an Ice cum freezing Plant, a well equipped processing unit for processing and marketing of diversified value added fish products, a research and development laboratory to carry out quality analysis of raw fish, finished products handled by the Project and testing and developing suitable packages for different value added products and Life Raft Servicing Station. The Institute has a unit at Vizag also.

Mandate of the Project

(a) Technology development and transfer to beneficiaries consisting of rural fishermen community, small scale industries and export processing houses through consultancy and job work.

(b) Value added product development by way of process and product diversification from all varieties of fish including low value and unconventional species.

(c) Imparting training in the field of Post Harvest Technology, Refrigeration Technology, Quality control and Value added Products.

(d) Providing consultancy and training for rural development programmes like supporting local fish farmers, self-help groups of fisher community and fishermen's co-operative societies functioning and Panchayati Raj Institutions. And also to Women empowerment programmes in processing and establishment of SSI units to self help groups.

(e) Popularization and test marketing of value added products from all fish varieties including low value and unconventional species.

(f) Extension of the popularization and test marketing of value added products to new areas and developing market in all states in a phased manner with added attention to rural areas enthusing entrepreneurs to enter into seafood processing industry.

(g) Function as a nodal agency on all matters associated with quality assurance, standardization of fish products and safety of fish products.

POSTMORTEM CHANGES IN FISH

Sensory Changes

Sensory changes are those perceived with the senses, i.e., appearance, odour, texture and taste.

Changes in Raw Fresh Fish

The first sensory changes of fish during storage are concerned with appearance and texture. The characteristic taste of the species is normally developed the first couple of days during storage in ice. The most dramatic change is onset of rigor mortis. Immediately after death the muscle is totally relaxed and the limp elastic texture usually persists for some hours, whereafter the muscle will contract. When it becomes hard and stiff the whole body becomes inflexible and the fish is in rigor mortis This condition usually lasts for a day or more and then rigor resolves. The resolution of rigor mortis makes the muscle relax again and it becomes limp, but no longer as elastic as before rigor. The rate in onset and resolution of rigor varies from species to species and is affected by temperature, handling, size and physical condition of the fish.

The effect of temperature on rigor is not uniform. In the case of cod, high temperatures give a fast onset and a very strong rigor mortis. This should be avoided as strong rigor tensions may cause gaping, i.e., weakening of the connective tissue and rupture of the fillet.

It has generally been accepted that the onset and duration of rigor mortis are more rapid at high temperatures, but observations, especially on tropical fish show the opposite effect of temperature with regard to the onset of rigor.

It is evident that in these species the onset of rigor is accelerated at 0°C compared to 10°C, which is in good correlation with a stimulation of biochemical changes at 0°C (Poulter *et al.,* 1982; Iwamoto *et al.,* 1987). However, an explanation for this has been suggested by Abe and Okuma (1991) who have shown that onset of rigor mortis in carp (Cyprinus *carpio)* depends on the difference in sea temperature and storage temperature. When the difference is large the time from death to onset of rigor is short and *vice versa.*

Rigor mortis starts immediately or shortly after death if the fish is starved and the glycogen reserves are depleted, or if the fish is stressed. The method used for stunning and killing the fish also influences the onset of rigor. Stunning and killing by hypothermia (the fish is killed in iced water) give the fastest onset of rigor, while a blow on the head gives a delay of up to 18 hours (Azam *et al.,* 1990; Proctor *et al.,* 1992).

The technological significance of rigor mortis is of major importance when the fish is filleted before or in rigor. In rigor the fish body will be completely stiff; the filleting yield will be very poor, and rough handling can cause gaping.

If the fillets are removed from the bone pre-rigor the muscle can contract freely and the fillets will shorten following the onset of rigor. Dark muscle may shrink up to 52 % and white muscle up to 15 % of the original length (Buttkus, 1963). If the fish is cooked pre-rigor the texture will be very soft and pasty.

In contrast, the texture is tough but not dry when the fish is cooked in rigor. Post-rigor the flesh will become firm, succulent and elastic.

Whole fish and fillets frozen pre-rigor can give good products if they are carefully thawed at a low temperature in order to give *rigor mortis* time to pass while the muscle is still frozen.

The sensory evaluation of raw fish in markets and landing sites is done by assessing the appearance, texture and odour. Most scoring systems are based upon changes taking place during storage in melting ice. It should be remembered that the characteristic changes vary depending on the storage method. The appearance of fish stored under chilled condition without ice does not change as much as for iced fish, but the fish spoil more rapidly and an evaluation of cooked flavour will be necessary. A knowledge of the time /temperature history of the fish should therefore be essential at landing.

The characteristic sensory changes in fish post mortem vary considerably depending on fish species and storage method. A general description has been provided by the EEC in the guidelines for quality assessment of fish. The suggested scale is numbered from 0 to 3, where 3 is the best quality.

The West European Fish Technologists' Association has compiled a multilingual glossary of odours and flavours which also can be very useful when looking for descriptive words for sensory evaluation of freshness of fish.

Changes in Eating Quality

If quality criteria of chilled fish during storing are needed, sensory assessment of the cooked fish can be conducted. A characteristic pattern of the deterioration of fish stored in ice can be found and divided into the following four phases:

- Phase 1 The fish is very fresh and has a sweet, seaweedy and delicate taste. The taste can be very slightly metallic. In cod, haddock, whiting and flounder, the sweet taste is maximized 2-3 days after catching.
- Phase 2 There is a loss of the characteristic odour and taste. The flesh becomes neutral but has no off-flavours. The texture is still pleasant.
- Phase 3 There is sign of spoilage and a range of volatile, unpleasant-smelling substances is produced depending on the fish species and type of spoilage (aerobic, anaerobic). One of the volatile compounds may be trimethylamine (TMA) derived from the bacterial

reduction of trimethyl-aminoxide (TMAO). TMA has a very characteristic "fishy" smell. At the beginning of the phase the off-flavour may be slightly sour, fruity and slightly bitter, especially in fatty fish. During the later stages sickly sweet, cabbage-like, ammoniacal, sulphurous and rancid smells develop. The texture becomes either soft and watery or tough and dry.

- Phase 4 The fish can be characterized as spoiled and putrid.

A numbered scale may be used for the sensory evaluation of cooked fish as shown in Figure.

The scale is numbered from 0 to 10, 10 indicating absolute freshness, 8 good quality and 6 a neutral tasteless fish.

The rejection level is 4. Using the scale in this way the graph becomes S-shaped indicating a fast degradation of the fish during the first phase, a slower rate in phase 2 and 3 and finally a high rate when the fish is spoiled.

Other scales can well be used and can change the shape of the graph. It is, however, important to understand the kind of results desired from the sensory analysis in order to ask the right questions to the sensory assessors.

Autolytic Changes

Autolysis means "self-digestion". It has been known for many years that there are at least two types of fish spoilage: bacterial and enzymatic. Uchyama and Ehira (1974) showed that for cod and yellowtail tuna, enzymatic changes related to fish freshness preceded and were unrelated to changes in the microbiological quality. In some species (squid, herring), the enzymatic changes precede and therefore predominate the spoilage of chilled fish. In others, autolysis contributes to varying degrees to the overall quality loss in addition to microbially-mediated processes.

Production of energy in post mortem muscle

At the point of death, the supply of oxygen to the muscle tissue is interrupted because the blood is no longer pumped by the heart and is not circulated through the gills where, in the living fish, it becomes enriched with oxygen. Since no oxygen is available for normal respiration, the production of energy from ingested nutrients is greatly restricted. Glycogen (stored carbohydrate) or fat is oxidized or "burned" by the tissue enzymes in a series of reactions which ultimately produce carbon dioxide (CO_2), water and the energy-rich organic compound adenosine triphosphate (ATP).

This type of respiration takes place in two stages: an anaerobic and an aerobic stage. The latter depends on the continued presence of oxygen (O_2) which is only available from the circulatory system. Most crustaceans are capable of respiring outside the aquatic environment by absorption of atmospheric oxygen for limited periods.

ATP may be synthesized by two other important pathways from creatine phosphate or from arginine phosphate. The former source of energy is restricted to vertebrate muscle (teleost fish) while the latter is characteristic of some invertebrates such as the cephalopods (squid and octopus). In either case, ATP production ceases when the creatine or arginine phosphates are depleted. It is interesting to note that octopine is the end-product from the anaerobic metabolism of cephalopods and is not acidic (unlike lactate), thus any changes in post mortem pH in such animals are not related to the production of lactic acid from glycogen.

For most teleost fish, glycolysis is the only possible pathway for the production of energy once the heart stops beating. This more inefficient process has principally lactic and pyruvic acids as its end-products. In addition, ATP is produced in glycolysis, but only 2 moles for each mole of glucose oxidized as compared to 36 moles ATP produced for each mole of glucose if the glycolytic end products are oxidized aerobically in the mitochondrion in the living animal. Thus, after death, the anaerobic muscle cannot maintain its normal level of ATP, and when the intracellular level declines from 7-10 µmoles/g to £ 1.0 µmoles/g tissue, the muscle enters rigor mortis. Post mortem glycolysis results in the accumulation of lactic acid which in turn lowers the pH of the muscle. In cod, the pH drops from 6.8 to an ultimate pH of 6.1-6.5. In some species of fish, the final pH may be lower: in large mackerel, the ultimate rigor pH may be as low as 5.8-6.0 and as low as 5.4-5.6 in tuna and halibut, however such low pH levels are unusual in marine teleosts.

These pHs are seldom as low as those observed for post mortem mammalian muscle. For example, beef muscle often drops to pH levels of 5.1 in rigor mortis. The amount of lactic acid produced is related to the amount of stored carbohydrate (glycogen) in the living tissue. In general, fish muscle contains a relatively low level of glycogen compared to mammals, thus far less lactic acid is generated after death.

Also, the nutritional status of the fish and the amount of stress and exercise encountered before death will have a dramatic effect on the levels of stored glycogen and consequently on the ultimate post mortem pH. As a rule, well-rested, well-fed fish contain more glycogen than exhausted fish. In a recent study of Japanese loach (Chiba et al., 1991), it was shown that only minutes of pre-capture stress resulted in a decrease of 0.50 pH units in 3 hours as compared to non-struggling fish whose pH dropped only 0.10 units in the same time period. In addition, the same authors showed that bleeding of fish significantly reduced the post mortem production of lactic acid.

The post mortem reduction in the pH of fish muscle has an effect on the physical properties of the muscle. As the pH drops, the net surface charge on the muscle proteins is reduced, causing them to partially denature and lose some of their water-holding capacity. Muscle tissue in the state of rigor mortis loses its moisture when cooked and is particularly unsuitable for further processing which involves heating, since heat denaturation enhances the water loss. Loss of water has a detrimental effect on the texture of fish muscle and it has been shown by Love (1975) that there is an inverse relationship between muscle toughness and pH, unacceptable levels of toughness (and water-loss on cooking) occurring at lower pH levels.

Autolysis and Nucleotide Catabolism

As mentioned earlier, rigor mortis sets in when the muscle ATP level drops to £ 1.0 µmoles/g. ATP is not only a source of high energy which is required for muscle contraction in the living animal, but also acts as a muscle plasticizer. Muscle contraction per se is controlled by calcium and an enzyme, ATP-ase which is found in every muscle cell.

When intracellular Ca^{+2} levels are 1 µM, Ca^{+2} - activated ATP-ase reduces the amount of free muscle ATP which results in the interaction between the major contractile proteins, actin and myosin. This ultimately results in the shortening of the muscle, making it stiff and inextensible. A fish in rigor mortis cannot normally be filleted or processed because the carcass is too stiff to be manipulated and is often contorted, making machine-handling impossible.

The resolution of rigor is a process still not completely understood but always results in the subsequent softening (relaxation) of the muscle tissue and is thought to be related to the activation of one or more of the naturally-occurring muscle enzymes, digesting away certain components of the rigor mortis complex.

The softening of the muscle during resolution of rigor (and eventually spoilage processes) is coincidental with the autolytic changes. Among the changes, one of the first to be recognized was the degradation of ATP-related compounds in a more-or-less predictable manner after death.

The degradation of ATP catabolites proceeds in the same manner with most fish but the speed of each individual reaction (from one catabolite to another) greatly varies from one species to another and often progresses coincidentally with the perceived level of spoilage as determined by trained analysts. Saito et al. (1959) were the first to observe this pattern and to develop a formula for fish freshness based on these autolytic changes:

$$K\% = \frac{[\text{Ino}]+[\text{Hx}]}{[\text{ATP}]+[\text{ADP}]+[\text{AMP}]+[\text{IMP}]+[\text{Ino}]+[\text{Hx}]} \times 100$$

where [ATP], [ADP], [AMP], [IMP], [Ino] and [Hx] represent the relative concentrations of these compounds in fish muscle measured at various times during chilled storage.

The K or "freshness" index gives a relative freshness rating based primarily on the autolytic changes which take place during post mortem storage of the muscle. Thus, the higher the K value, the lower the freshness level. Unfortunately, some fish species such as Atlantic cod reach a maximum K value well in advance of the shelf life as determined by trained judges, and K is therefore not considered reliable as a freshness index for all marine finfish.

Also, the degradation of nucleotide catabolites is only coincidental with perceived changes in freshness and not necessarily related to the cause of freshness deterioration since only Hx is considered to have a direct effect on the perceived bitter off-flavour of spoiled fish (Hughes and Jones, 1966). It is now widely accepted that IMP is responsible for the desirable fresh fish flavour which is only present in top quality seafood. None of the nucleotide catabolites are considered to be related to the perceived changes in texture during the autolytic process except of course ATP whose loss is associated with *rigor mortis*.

Surette *et al.* (1988) followed the autolysis of sterile and non-sterile cod as indicated by the ATP catabolites. The rates of formation and breakdown of IMP were the same in both sterile and non- sterile samples of cod tissue, indicating that the catabolic pathway for the degradation of ATP through to inosine is entirely due to autolytic enzymes.

The conversion of Iino to Hex was accelerated by about 2 days for the non-sterile samples, suggesting that bacterial nucleoside phosphorylase plays a major role in the *postmortem* production of Hx in refrigerated cod. It is interesting to note that Surette *et al.* (1988) were not able to recover nucleoside phosphorylase from freshly killed cod, but Surette *et al.* (1990) later went on to isolate and purify this enzyme from a *Proteus* bacterium recovered from spoiled cod fillets. As mentioned earlier, large variations can be expected in the patterns of nucleotide degradation from one species to another.

There is little doubt that physical handling accelerates the autolytic changes in chilled fish. Surette et al. (1988) reported that the breakdown rate of the nucleotide catabolites was greater in sterile fillets than in non-sterile gutted whole cod. This is perhaps not surprising since many of the autolytic enzymes have been shown to be compartmentalized in discrete membrane-bound packages which become broken when subjected to physical abuse and result in the intimate mixing of enzyme and substrate. Crushing of the fish by ice or other fish can seriously affect the edibility and filleting yields even for fish which have a relatively low bacterial load, demonstrating the importance of autolytic processes. Iced fish should never be stored in boxes deeper than 30 cm and it is equally important to be sure that fish boxes are not permitted to "nest" one on top of the other if autolysis is to be minimized. Systems for conveying fish and for discharge from the vessels must be designed so as to avoid physical damage to the delicate tissues.

Several rapid methods have been developed for the determination of individual nucleotide catabolites or combinations including the freshness index. Two recent reviews should be consulted.

Autolytic Changes Involving Proteolytic Enzymes

Many proteases have been isolated from fish muscle and the effects of proteolytic breakdown are often related to extensive softening of the tissue. Perhaps one of the most notable examples of autolytic proteolysis is the incidence of belly-bursting in pelagic (fatty fish) species such as herring and capelin. This type of tissue softening is most predominant in summer months when pelagics are feeding heavily, particularly on "red feed" consisting of copepods and euphausiids. The low molecular weight peptides and free amino-acids produced by the autolysis of proteins not only lower the commercial acceptability of pelagics, but in bulk-stored capelin, autolysis has been shown to accelerate the growth of spoilage bacteria by providing a superior growth environment for such organisms (Aksnes and Brekken, 1988). The induction of bacterial spoilage in capelin by autolysis also resulted in the decarboxylation of amino-acids, producing biogenic amines and lowered the nutritive value of the fish significantly. This is particularly important since autolysis and bacterial growth greatly lower the commercial value of pelagics used for the production of fishmeal.

Similarly, bulk-stored herring used for fishmeal has been found to contain carboxy-peptidases A and B, chymotrypsin, and trypsin; and preliminary studies have shown that proteolysis can be inhibited by the addition of potato extracts which not only slowed the proteolysis but resulted in lower microbial growth and preservation of the nutritional value of the meal (Aksnes, 1989).

More recently, Botta *et al.* (1992) found that autolysis of the visceral cavity (belly-bursting) of herring was related more to physical handling practices than to biological factors such as fish size, amount of red feed in the gut or roe content. In particular, it was shown that for herring, freezing/thawing, thawing time at 15°C and time of iced storage, had a far greater influence on belly- bursting than biological factors.

Cathepsins

Although several proteolytic enzymes have been discovered in the fish tissues, it has perhaps been the cathepsins which have been described most often. The cathepsins are "acid" proteases usually found packaged in tiny, submicroscopic organelles called lysozomes. In living tissue, lysozomal proteases are believed to be responsible for protein breakdown at sites of injury. Thus cathepsins are for the most part inactive in living tissue but become released into the cell juices upon physical abuse or upon freezing and thawing of *post mortem* muscle.

Cathepsins D and L are believed to play a major role in the autolytic degradation of fish tissue since most of the other cathepsins have a relatively narrow pH range of activity far too low to be of physiological significance. Reddi *et al.* (1972) demonstrated that an enzyme believed to be cathepsin D from winter flounder was active over a pH range of 3-8 with a maximum near pH 4.0, although no attempt was made to confirm the identity of the enzyme using synthetic substrates or specific inhibitors. Nevertheless, the enzyme was far less active in the presence of ATP, suggesting that such an enzyme would only be active in *post mortem* fish muscle. Also, the enzyme activity was inhibited strongly by the presence of salt with virtually no activity remaining after a 25-hour incubation in the presence of 5% sodium chloride. It is therefore unlikely that Reddi's enzyme was active in salted fish products.

Cathepsin L has been implicated in the softening of salmon muscle during spawning migration. It is likely that this enzyme contributes more to autolysis of fish muscle than cathepsin D since it is far more active at neutral pH, and has been shown to digest both myofibrillar proteins (actomyosin) as well as connective tissue. Yamashita and Konogaya (1990) produced strong evidence implicating cathepsin L rather than other cathepsins in the softening of salmon during spawning.

They demonstrated that electrophoresis of purified myofibrils treated with cathepsin L resulted in patterns which were almost identical to patterns of proteins recovered from muscle from spawning fish. Furthermore, the cathepsin L autolytic activity correlated well with the texture of the muscle as measured instrumentally. The linear correlation between cathepsin L activity and breaking strength of the muscle was excellent; r = 0.86 and -0.95 for fresh and frozen/thawed tissue, respectively.

It is interesting that, in all cases, the autolytic ability as measured by cathepsin L activity was higher in frozen/thawed tissue than in fresh tissue. Freezing and thawing often break down cell membranes allowing autolytic membrane-bound enzymes to react with their natural substrates. The enzyme and its naturally occurring inhibitor were further studied by the same authors (Yamashita and Konogaya, 1992). Cathepsin L has also been associated with the production of a jelly- like softening of flounder (Toyohara *et l.*, 1993 a) and the uncontrollable softening of Pacific hake muscle which has been parasitized by Myxosporidia (Toyohara *et al.*, 1993 b). The tissues of such infected fish have little commercial value, but at present it is not known if it is the parasite or the host which secretes the proteolytic enzymes which autolyze the muscle. In addition to their detrimental effect on texture, catheptic enzymes induce intentional autolytic changes in fermented fish products. For example, cathepsins are believed to be responsible for major textural changes during the fermentation of salted preserved Japanese squid and Crucian carp (Makinodan *et al.*, 1991, 1993).

Calpains

A second group of intracellular proteases called "calpains" or "calcium activated factor" (CAF) has recently been associated with fish muscle autolysis and is found in meats, finfish and crustaceans. Tenderness is probably the most important quality characteristic of red meat.

It has been known for nearly a century that *post mortem* aging of red meat results in the tenderization process. Calpains have been found primarily responsible for the *post mortem* autolysis of meat through digestion of the z- line proteins of the myofibril.

Although toughness is seldom a problem with unfrozen fish muscle, softening through autolysis is a serious problem limiting the commercial value. The calpains are intra-cellular endopeptidases requiring cysteine and calcium; μ-calpain requiring 5-50 μM Ca^{2+}, m-calpain requiring 150-1000 μM Ca^{2+}. Most calpains are active at physiological pH, making it reasonable to suspect their importance in fish-softening during chilled storage.

Studies have shown that in crustacean muscle, calpains are associated with moltinduced textural changes to the muscle and carry out non-specific generalized digestion of the myofibrillar proteins.

However, vertebrate muscle calpains have been shown to be very specific, digesting primarily tropinin- T, desmin, titin and nebulin, attacking neither vertebrate actin or myosin (Koohmaraie, 1992). In contrast, fish calpains digest myosin (specifically the myosin heavy chain) to form an initial fragment with approximate molecular weight of 150 000 Da (Muramoto *et al.*, 1989).

The same authors demonstrated that fish calpains were far more active at low temperatures than were mammalian calpains and that the rates of cleavage were species-specific, being most active against myosins with lowest heat stabilities. Thus, fish species adapted to colder environmental temperatures are more susceptible to calpain autolysis than those from tropical waters.

Although calpain has been identified in several fish species including carp (Toyohara *et al.*, 1985), tilapia and shrimp (Wang *et al.*, 1993), as well as tuna, croaker, red seabream and trout (Muramoto *et al.*, 1989) to name a few, little work has to date demonstrated a "cause and effect" relationship between calpain activity and instrumental measurements of texture.

Collagenases

To this point, all of the *post mortem* autolytic changes described have involved changes within the muscle cell *per se*. However, the flesh of teleost fish is divided into blocks of muscle cells separated into "flakes" or myotomes by connective tissue called myocommata.

Each muscle cell or fibre is surrounded with connective tissue which attaches to the myocommata at the ends of the cells by means of fine collagenous fibrils. During chilled storage, these fibrils deteriorate.

More recently, it was shown that instrumental measurements of texture of chilled trout muscle decreased as the amount of type V collagen was solubilized, presumably due to the action of autolytic collagenase enzymes. It is these enzymes which presumably cause "gaping" or breakdown of the myotome during long-term storage on ice or short term storage at high temperature.

For Atlantic cod, it has been shown that upon reaching 17°C, gaping is inevitable presumably because of degradation of the connective tissue and rapid shortening of the muscle due to high temperature *rigor*.

The relatively short shelf life of chilled prawns due to softening of the tissue has also been shown to be due to the presence of collagenase enzymes (Nip *et al.*, 1985). The source of the collagenase enzymes in prawn is thought to be the hepatopancreas (digestive organ).

Autolytic Changes during Frozen Storage

The reduction of trimethylamine oxide (TMAO), an osmoregulatory compound in many marine teleost fish, is usually due to bacterial action but in some species an enzyme is present in the muscle tissue which is able to break down TMAO into dimethylamine (DMA) and formaldehyde (FA):

$$(CH_3)_3\,NO \rightarrow (CH_3)_2NH + HCHO$$

It is important to note that the amount of formaldehyde produced is equivalent to the dimethylamine formed but is of far greater commercial significance. Formaldehyde induces cross- linking of the muscle proteins making the muscle tough and readily lose its water holding capacity.

The enzyme responsible for formal dehyde-induced toughening is called TMAO-ase or TMAO demethylase and is most commonly found in the gadoid fishes (cod family). Most of the TMAO demethylase enzymes reported to date weremembrane-bound and become most active when the tissue membranes are disrupted by freezing or artificially by detergent solubilization.

Dark (red) muscle has a higher rate of activity than white muscle whereas other tissues such as kidney, spleen and gall bladder are extremely rich in the enzyme. Thus, it is important that minced fish is completely free of organ tissue such as kidney from gadoid species if toughening in frozen storage is to be avoided.

It is often difficult to ensure that the kidney is removed prior to mechanical deboning since this particular organ runs the full length of the backbone and is adherent to it. The TMAO-ase enzyme has been isolated from the microsomal fraction in hake muscle (Parkin and Hultin, 1986) and the lysosomal membrane in kidney tissue (Gill *et al.*, 1992). It has been shown that the toughening of frozen hake muscle is correlated to the amount of formaldehyde produced, and that the rate of FA production is greatest at high frozen-storage temperatures (Gill *et al., 1979).*

In addition, it has been shown that the amount of FA-induced toughening is enhanced by physical abuse to the catch prior to freezing and by temperature fluctuations during frozen storage. The most practical means of preventing the autolytic production of FA is to store fish at temperatures < –30°C to minimize temperature fluctuations in the cold store and to avoid rough handling or the application of physical pressure on the fish prior to freezing.

Generally, the most important single factor affecting autolysis is physical disruption of the muscle cells. No attempt has been made here to deal with the alkaline proteases associated with the softening of cooked surimi products. An article by Kinoshita *et al.* (1990) deals with the heat-activated alkaline proteases associated with the softening in surimi-based products.

BACTERIOLOGICAL CHANGES

The Bacterial Flora on Live Fish

Microorganisms are found on all the outer surfaces (skin and gills) and in the intestines of live and newly caught fish. The total number of organisms vary enormously and Liston (1980) states a normal range of 10^2–10^7 cfu (colony forming units)/cm^2 on the skin surface. The gills and the intestines both contain between 10^3 and 10^9 cfu/g (Shewan, 1962).

The bacterial flora on newly-caught fish depends on the environment in which it is caught rather than on the fish species (Shewan, 1977). Fish caught in very cold, clean waters carry the lower numbers whereas fish caught in warm waters have slightly higher counts. Very high numbers, i.e., 10^7 cfu/cm^2 are found on fish from polluted warm waters. Many different bacterial species can be found on the fish surfaces. The bacteria on temperate water fish are all classified according to their growth temperature range as either psychrotrophs or psychrophiles. Psychrotrophs (cold-tolerant) are bacteria capable of growth at 0°C but with optimum around 25°C.

Psychrophiles (cold-loving) are bacteria with maximum growth temperature around 20°C and optimum temperature at 15°C (Morita, 1975). In warmer waters, higher numbers of mesophiles can be isolated.

The microflora on temperate water fish is dominated by psychrotrophic Gram-negative rodshaped bacteria belonging to the genera *Pseudomonas, Moraxella, Acinetobacter, Shewanella* and *Flavobacterium.*

Members of the *Vibrionaceae (Vibrio* and *Photobacterium)* and the *Aeromonadaceae (Aeromonas* spp.) are also common aquatic bacteria and typical of the fish flora. Gram-positive organisms as *Bacillus, Micrococcus, Clostridium, Lactobacillus* and coryneforms can also be found in varying proportions, but in general, Gram-negative bacteria dominate the microflora. Shewan (1977) concluded that Gram-positive *Bacillus* and Micrococcus dominate on fish from tropical waters.

However, this conclusion has later been challenged by several studies which have found that the microflora on tropical fish species is very similar to the flora on temperate species (Acuff *et al.,*1984; Gram *et al.,* 1990; Lima dos Santos 1978; Surendran *et al.,* 1989).

A microflora consisting of *Pseudomonas, Acinetobacter, Moraxella* and *Vibrio* has been found on newly-caught fish in several Indian studies (Surendran *et al.,* 1989). Several authors conclude, as Liston (1980), that the microflora on tropical fish often carry a slightly higher load of Gram-positives and enteric bacteria but otherwise is similar to the flora on temperate-water fish.

Aeromonas spp. are typical of freshwater fish, whereas a number of bacteria require sodium for growth and are thus typical of marine waters. These include *Vibrio, Photobacterium* and *Shewanella.* However, although *Shewanella putrefaciens* is characterized as sodium-requiring, strains of *S. putrefaciens* can also be isolated from freshwater environments (DiChristina and DeLong, 1993; Gram et al., 1990; Spanggaard et al., 1993). Although *S. putrefaciens* has been isolated from tropical freshwaters, it is not important in the spoilage of freshwater fish (Lima dos Santos, 1978; Gram, 1990).

Table: Bacterial flora on fish caught in clean, unpolluted waters

Gram-negative	Gram-positive	Comments
Pseudomonas	*Bacillus*	
Moraxella	*Clostridium*	
Acinetobacter	*Micrococcus*	
Shewanella putrefaciens	*Lactobacillus*	
Flavobacterium	Coryneforms	
Cytophaga		
Vibrio Photobacterium *Aeromonas*		*Vibrio* and *Photobacterium* are typical of marine waters; *Aeromonas* is typical of freshwater

In polluted waters, high numbers of *Enterobacteriaceae* may be found. In clean temperate waters, these organisms disappear rapidly, but it has been shown that *Escherichia coli* and *Salmonella* can survive for very long periods in tropical waters and once introduced may almost become indigenous to the environment (Fujioka *et al.,* 1988).

The taxonomy of S. *putrefaciens* has been rather confused. The organism was originally associated with the *Achromobacter* group but was later placed in the Shewan *Pseudomonas* group IV.

Based on percentage of guanine+ cytosine (GC%) it was transferred to the genus *Alteromonas,* but on the basis of 5SRNA homology it was reclassified to a new genus, *Shewanella* (MacDonnell and Colwell, 1985). It has recently been suggested that the genus *Aeromonas* spp. which was a member of the *Vibrionaceae* family be transferred to its own family, the *Aeromonadaceae* (Colwell *et al.,* 1986).

Japanese studies have shown very high numbers of microorganisms in the gastrointestinal tract of fish, and as such numbers are much higher than in the surrounding water, this indicates the presence of a favourable ecological niche for the microorganisms.

Similarly, Larsen *et al.* (1978) reported up to 10^7 cfu/g of vibrio-like organisms in the intestinal tract of cod and Westerdahl *et al.* (1991) also isolated high numbers of vibrio-like organisms from the intestines of turbot. *Photobacterium phosphoreum* which can be isolated from the surface can also be isolated in high numbers from the intestinal tract of some fish species (Dalgaard, 1993).

On the contrary, some authors believe that the microflora of the gastrointestinal tract is merely a reflection of the environment and the food intake.

Microbial Invasion

The flesh of healthy live or newly-caught fish is sterile as the immune system of the fish prevents the bacteria from growing in the flesh. When the fish dies, the immune system collapses and bacteria are allowed to proliferate freely. On the skin surface, the bacteria to a large extent colonize the scale pockets. During storage, they invade the flesh by moving between the muscle fibres. Murray and Shewan (1979) found that only a very limited number of bacteria invaded the flesh during iced storage.

Ruskol and Bendsen (1992) showed that bacteria can be detected by microscope in the flesh when the number of organisms on the skin surface increases above 10^6 cfu/cm^2. This was seen at both iced and ambient temperatures. No difference was found in the invasive patterns of specific spoilage bacteria (e.g., S. *putrefaciens)* and non-spoilage bacteria.

Since only a limited number of organisms actually invade the flesh and microbial growth mainly takes place at the surface, spoilage is probably to a large extent a consequence of bacterial enzymes diffusing into the flesh and nutrients diffusing to the outside.

Fish spoil at very different rates, and differences in surface properties of fish have been proposed to explain this. Skins of fish have very different textures. Thus whiting *(Merlangius merlangus)* and cod *(Gadus morhua)* which have a very fragile integument spoil rapidly compared to several flatfish such as plaice that has a very robust dermis and epidermis. Furthermore, the latter group has a very thick slime layer, which includes several antibacterial components, such as antibodies, complement and bacteriolytic enzymes (Murray and Fletcher, 1976; Hjelmland *et al.,* 1983).

Changes in the Microflora during Storage and Spoilage/ Specific Spoilage Organisms

Bacteria on fish caught in temperate waters will enter the exponential growth phase almost immediately after the fish have died. This is also true when the fish are iced, probably because the microflora is already adapted to the chill temperatures. During ice storage, the bacteria will grow with a doubling time of approximately 1 day and will, after 2-3 weeks, reach numbers of 10^8-10^9 cfu/g flesh or cm^2 skin. During ambient storage, a slightly lower level of 10^7-10^8 cfu/g is reached in 24 hours. The bacteria on fish caught in tropical waters will often pass through a lag-phase of 1-2 weeks if the fish are stored in ice, whereafter exponential growth begins. At spoilage, the bacterial level on tropical fish is similar to the levels found on temperate fish species.

If iced fish are stored under anaerobic conditions or if stored in CO_2 containing atmosphere, the number of the normal psychrotrophic bacteria such as S. *putrefaciens* and *Pseudomonas* is often much lower, i.e., 10^6-10^7 cfu/g than on the aerobically stored fish. However, the level of bacteria of psychrophilic character such as P. *phosphoreum* reaches a level of 10^7-10^8 cfu/g when the fish spoil.

The composition of the microflora also changes quite dramatically during storage. Thus, under aerobic iced storage, the flora is composed almost exclusively of *Pseudomonas* spp. and S. *putrefaciens* after 1-2 weeks. This is believed to be due to their relatively short generation time at chill temperatures and is true for all studies carried out whether on tropical or temperate-water fish. At ambient temperature (25°C), the microflora at the point of spoilage is dominated by mesophilic *Vibrionaceae* and, particularly if the fish are caught in polluted waters, *Enterobacteriaceae.*

A clear distinction should be made between the terms spoilage flora and spoilage bacteria since the first describes merely the bacteria present on the fish when it spoils whereas the latter is the specific group that produce the off-odours and off-flavours associated with spoilage. A large part of the bacteria present on the spoiled fish have played no role whatever in the spoilage. Each fish product will have its own specific spoilage bacteria and the number of these will, as opposed to the total number, be related to the shelf life.

It is not an easy task to determine which of the bacteria isolated from the spoiled fish are those causing spoilage, and it requires extensive sensory, microbiological and chemical studies. First, the sensory, microbiological and chemical changes during storage must be studied and quantified, including a determination of the level of a given chemical compound that correlates with spoilage (the chemical spoilage indicator). Second, bacteria are isolated at the point of sensory rejection. Pure and mixed cultures of bacteria are screened in sterile fish substrates for their spoilage potential, i.e., their ability to produce sensory (off-odours) and chemical changes typical of the spoiling product. Finally, the selected strains are tested to evaluate their spoilage activity, i.e., if their growth rate and their qualitative and quantitative production of off-odours are similar to the measurements in the spoiled product.

The latter step is particularly important, as some bacteria may produce the chemical compounds associated with spoilage but are unable to do so in significant amounts, and they are thus not the specific spoilage bacteria. When stored aerobically, levels of 10^8-10^9 cfu/g of specific spoilage bacteria are required to cause spoilage. The spoilage of packed fish is seen at a much lower level of 10^7 cfu P. *phosphoreum* per gramme. This relatively low level is probably due to the very large size (5 μm) of the bacterium resulting in a much higher yield of for example,

TMA per cell (Dalgaard, 1993). Spoilage potential and activity can be assessed in several fish substrates as sterile, raw fish juice (Lerke *et al.*, 1963), heat-sterilized fish juice (Castell and Greenough, 1957; Gram *et al.* , *1987*; Dalgaard, 1993) or on sterile muscle blocks (Herbert *et al.*, 1971). The latter is the most complicated but is also that yielding results comparable to the product. If any of the fish juices are chosen, it is important that the growth rate of the spoilage bacteria in the model system is equal to the growth rate in the product.

A qualitative test for the ability of the bacteria to produce H_2S and/or reduce TMAO may also be used when the spoilage flora is screened for potential spoilage bacteria. A medium where the reduction of TMAO to TMA is seen as a redox indicator changes colour, and the formation of H_2S is evident from a black precipitation of FeS which has been developed for this purpose (Gram *et al.*,1987).

Shewanella putrefaciens has been identified as the specific spoilage bacteria of marine temperate- water fish stored aerobically in ice. If the product is vacuum-packed, P. phosphoreum participates in the spoilage and it becomes the specific spoilage bacteria of CO_2 packed fish. The spoilage flora on iced tropical fish from marine waters is composed almost exclusively of *Pseudomonas* spp. and *S. putrefaciens*. Some *Pseudomonas* spp. are the specific spoilers of iced stored tropical freshwater fish (Lima dos Santos, 1978; Gram et al., 1990) and are also, together with *S. putrefaciens*, spoilers of marine tropical fish stored in ice (Gillespie and MacRae, 1975; Gram, 1990).

At ambient temperature, motile aeromonads are the specific spoilers of aerobically stored freshwater fish (Gorzyka and Pek Poh Len, 1985; Gram *et al.*, 1990). Barile *et al.* (1985) showed that a large proportion of the flora on ambient-stored mackerel consisted of *S. putrefaciens,* indicating that this bacterium may also take part in the spoilage.

(1) Modified Atmosphere Packaging (CO_2 containing)

(2) LAB: Lactic Acid Bacteria

(3) Fish caught in tropical waters or freshwaters tend to have a spoilage dominated by *Pseudomonas spp.*

Biochemical Changes Induced by Bacterial Growth during Storage and Spoilage

These include trimethylamine, volatile sulphur compounds, aldehydes, ketones, esters, hypoxanthine as well as other low molecularweight compounds.

The substrates for the production of volatiles are the carbohydrates (e.g., lactate and ribose), nucleotides (e.g., inosine mono-phosphate and inosine) and other NPN molecules. The amino-acids are particularly important substrates for formation of sulphides and ammonia.

Microorganisms obtain far more energy from aerobic oxidation than from an anaerobic fermentation; thus the complete oxidation of 1 mole glucose (or other hexose) via Kreb's cycles yields 6 moles of CO_2 and 36 moles of ATP. On the contrary, the fermentation of 1 mole glucose gives only 2 moles of ATP and two moles of lactic acid. The initial aerobic growth on fish is dominated by bacteria using carbohydrates as substrate and oxygen as terminal electron-acceptor with the concurrent production of CO_2 and H_2O.

Reduction of Trimethylarnine Oxide (TMAO)

The growth of oxygen-consuming bacteria results in the formation of anaerobic or microaerophilic niches on the fish. This does, however, not necessarily favour the growth of anaerobic bacteria. Some of the bacteria present on fish are able to carry out a respiration (with the ATP advantage) by using other molecules as electron acceptor. It is typical of many of the specific spoilage bacteria on fish that they can use TMAO as electron acceptorin an anaerobic respiration. The reduced component, TMA, which is one of the dominant components of spoiling fish, has a typical fishy odour. The level of TMA found in fresh fish rejected by sensory panels varies between fish species, but is typically around 10-15 mg TMA-N/100 g in aerobically stored fish and at a level of 30 mg TMA-N/100 g in packed cod.

The TMAO reduction is mainly associated with the genera of bacteria typical of the marine environment *(Alteromonas,*

Photobactetium, Vibrio and S. *putrefaciens),* but is also carried out by *Aeromonas* and intestinal bacteria of the *Enterobacteriaceae.* TMAO reduction has been studied in fermentative, facultative anaerobic bacteria like *E. coli* (Sakaguchi *et al.,* 1980) and *Proteus* spp. (Stenberg *et al.,* 1982) as well as in the non-fermentative S. *putrefaciens* (Easter *et al,* 1983; Ringo *et al,* 1984).

During aerobic growth, S. *putrefaciens* uses the Kreb's cycle to produce the electrons that are later channelled through the respiratory chain. Ringo *et al.* (1984) suggested that during anaerobic respiration S. *putrefaciens* also uses the complete Kreb's cycle, whereas it has recently been shown that in the anaerobic respiration in S. *putrefaciens,* only part of the Kreb's cycle is used and electrons are also generated by another metabolic pathway, namely the serine pathway (Scott and Nealson, 1994). S. *putrefaciens* can use a variety of carbon sources as substrate in its TMAO-dependent anaerobic respiration, including formate and lactate. Compounds like acetate and succinate that are used in the oxygen respiration cannot be used when TMAO is terminal electron acceptor (DiChristina and DeLong, 1994) and on the contrary, acetate is a product of the anaerobic TMAO reduction (Ringo *et al.,* 1984; Scott and Nealson, 1994).

Contrary to this, sugars and lactate are the main substrates generating electrons when *Proteus* spp. reduces TMAO. The reduction is accompanied by a production of acetate as the main product (Kjosbakken and Larsen, 1974).

TMAO is, a typical component of marine fish, and it has recently been reported that also some tropical freshwater fish contain high amounts of TMAO (Anthoni *et al.,* 1990). However, TMA is not necessarily a characteristic component during spoilage of such fish because spoilage is due to *Pseudomonas spp.* (Gram *et al.,* 1990).

The development of TMA is in many fish species paralleled by a production of hypoxanthine. Hypoxanthine can be formed by the autolytic decomposition of nucleotides, but it can also be formed by bacteria; and the rate of bacterial formation is higher than the autolytic. Both Jorgensen *et al.* (1988) and Dalgaard (1993) showed a linear correlation between the contents of TMA and hypoxanthine during iced storage of packed cod. Several of the spoilage bacteria produce hypoxanthine from inosine or inosine mono-phosphate, including *Pseudomonas* spp. (Surette *et al.*, 1988) S. *putrefaciens* (van Spreekens, 1977; Jorgensen and Huss, 1989; Gram, 1989) and P. *phosphoreum* (van Spreekens, 1977).

In cod and other gadoid fishes, TMA constitutes most of the so-called total volatile bases, TVB (also called total volatile nitrogen, TVN) until spoilage. However, in the spoiled fish where the TMAO supplies are depleted and TMA has reached its maximum level, TVB levels still rise due to formation of NH3 and other volatile amines. A little ammonia is also formed in the first weeks of iced storage due to autolysis. In some fish that do not contain TMAO or where spoilage is due to a non-TMAO reducing flora, a slow rise in TVB is seen during storage, probably resulting from the deamination of amino-acids.

Volatile sulphur-compounds are typical components of spoiling fish and most bacteria identified as specific spoilage bacteria produce one or several volatile sulphides. S. *putrefaciens* and some *Vibrionaceae* produce H_2S from the sulphur containing amino-acid 1-cysteine (Stenstroem and Molin, 1990; Gram *et al.*, 1987). On the contrary, neither *Pseudomonas* nor P. *phosphoreum* produce significant amounts of H_2S. Thus, hydrogen sulphide, which is typical of spoiling iced cod stored aerobically, is not produced in spoiling CO_2 packed fish (Dalgaard *et al.*, 1993).

Methylmercaptan (CH_3SH) and dimethylsulphide ($(CH_3)_2S$) are both formed from the other sulphur-containing amino-acid, methionine. Taurine, which is also sulphur-containing, occurs as free amino-acid in very high concentrations in fish muscle. It disappears from the fish flesh during storage but this is because of leakage rather than because of bacterial attack (Herbert and Shewan, 1975).

The volatile sulphur-compounds are very foul-smelling and can be detected even at ppb levels, so even minimal quantities have a considerable effect on quality. Ringo *et al.* *(1984)* have shown that cysteine is used as substrate in the Kreb's cycle when electrons are transferred to TMAO, and the formation of H_2S and TMA is thus to some extent a linked reaction. Contrary to the iced spoilage by S. *putrefaciens* and the ambient spoilage by *Vibrionaceae* which is dominated by H_2S and TMA, the spoilage caused by *Pseudomonas* spp. is characterized by absence of these compounds (Gram *et al.*, 1989, Gram *et al.*, 1990).

Table : Typical spoilage compounds during spoilage of fresh fish stored aerobically or packed in ice or at ambient temperature

Specific spoilage organism	Typical spoilage compounds
Shewanella putrefaciens	TMA, H_2S, CH_3SH, $(CH_3)_2S$, Hx
Photobacterium phosphoreum	TMA, Hx
Pseudomonas spp.	ketones, aldehydes, esters, non-H_2S sulphides
Vibrionaceae	TMA, H_2S
anaerobic spoilers	NH_3, acetic, butyric and propionic acid

Fruity, rotten, sulphydryl odours and flavours are typical of the Pseudomonas spoilage of iced fish. *Pseudomonas* spp. produce a number of volatile aldehydes, ketones, esters and sulphides (Edwards *et al.*, 1987; Miller *et al.*, 1973 a, 1973 b). However, it is not known which specific compounds are responsible for the typical off odours. The fruity off-odours produced by *Pseudomonas fragi* originate from monoaminomonocarboxylic amino acids.

As mentioned above, TVB will continue to rise even after TMA has reached its maximum. This latter rise is due to proteolysis commencing when several of the free amino-acids have been used. Lerke *et al.* (1967) separated fish juice into a protein and a non-protein fraction and inoculated spoilage bacteria in each fraction and in the whole juice. The non-protein fraction of a fish juice spoiled as the whole juice whereas only faint off-odours were detected in the protein fraction of the juice. Although some authors have used the number of proteolytic bacteria as indicators of spoilage, it must be concluded that the turnover of the protein fraction is not of major importance in spoilage of fresh fish.

Some of the compounds typically formed by bacteria during spoilage of fish are shown in Table below together with the substrate used for the formation. The formation of TMA is accompanied by a formation of ammonia during anaerobic storage of herring and mackerel (Haaland and Njaa, 1988). Prolonged anaerobic storage of fish results in vigorous production of NH_3 owing to further degradation of the amino-acids, and in the accumulation of lower fatty acids as acetic, butyric and propionic acid.

The very strong NH_3-producers were found to be obligate anaerobes belonging to the family Bacteroidaceae genus *Fusobacterium* (Kjosbakken and Larsen, 1974; Storroe *et al.*, 1975, 1977). These organisms grow only in the spoiled fish extract and have little or no proteolytic activity relying on already hydrolysed proteins.

Table : Substrate and off-odour/off-flavour compounds produced by bacteria during spoilage of fish

Substrate	Compounds produced by bacterial action
TMAO	TMA
cysteine	H_2S
methionine	CH_3SH, $(CH_3)_2S$
carbohydrates and lactate	acetate, CO_2, H_2O
inosine, IMP	hypoxanthine
amino-acid s (glycine, serine, leucine)	esters, ketones, aldehydes
amino-acids, urea	NH_3

During iced storage of fresh fatty fish, changes in the lipid fraction is caused almost exclusively by chemical action, e.g., oxidation, whereas bacterial attack on the lipid fraction contributes little to the spoilage profile. During storage of lightly preserved fish, lipid hydrolysis caused by bacteria may be part of the spoilage profile.

LIPID OXIDATION AND HYDROLYSIS

The two distinct reactions in fish lipids of importance for quality deterioration are:

- oxidation
- hydrolysis

They result in production of a range of substances among which some have unpleasant (rancid) taste and smell. Some may also contribute to texture changes by binding covalently to fish muscle proteins. The various reactions are either *nonenzymatic* or catalyzed by *microbial* enzymes or by *intracellular* or *digestive* enzymes from the fish themselves. The relative significance of these reactions, therefore, mainly depends on fish species and storage temperature.

Fatty fish are, of course, particularly susceptible to lipid degradation which can create severe quality problems even on storage at subzero temperatures.

Oxidation

The large amount of polyunsaturated fatty acid moieties found in fish lipids makes them highly susceptible to oxidation by an autocatalytic mechanism. The process is initiated as described below by abstraction of a hydrogen atom from the central carbon of the *pentadiene structure* found in most fatty acid acyl chains containing more than one double bond:

$$-CH = CH-CH_2-CH = CH- \rightarrow -CH = CH-\dot{C}H-CH = CH- + H$$

Contrary to the native molecule, the lipid radical (L) reacts very quickly with atmospheric oxygen making a peroxy-radical (LOO) which again may abstract a hydrogen from another acyl chain resulting in a lipid hydroperoxide (LOOH) and a new radical L. This propagation continues until one of the radicals is removed by reaction with another radical or with an *antioxidant* (AH) whose resulting radical (A) is much less reactive. The hydroperoxides produced in relatively large amounts during propagation are tasteless, and it is therefore perhaps not surprising that the widely used "peroxide value" usually correlates rather poorly to sensorial properties.

The hydroperoxides are readily broken down, catalyzed by heavy metal ions, to secondary autoxidation products of shorter carbon chain-length. These secondary products - mostly aldehydes, ketones, alcohols, small carboxylic acids and alkanes - give rise to a very broad odour spectrum and in some cases to a yellowish discoloration. Several of the aldehydes can be determined as "thiobarbituric acid-reactive substances".

Metal ions are very important in the first step of lipid autoxidation - the initiation process - in catalyzing the formation of reactive oxygen species as for example the hydroxyl radical (OH).

This radical immediately reacts with lipids or other molecules at the site where it is generated. The high reactivity may explain that free fatty acids have been found to be more susceptible to oxidation than the corresponding bound ones, because the amount of iron in the aqueous phase is probably greater than the amount bound to the surface of cellular membranes and lipid droplets.

Fatty acid hydroperoxides may also be formed enzymatically, catalyzed by *lipoxygenase* which is present in variable amounts in different fish tissues. A relatively high activity has been found in the gills and under the skin of many species. The enzyme is unstable and is probably important for lipid oxidation only in fresh fish. Cooking or freezing/thawing rather effectively destroys the enzyme activity.

The living cells possess several protection mechanisms directed against lipid oxidation products. An enzyme, glutathione peroxidase, exists which reduces hydroperoxides in the cellular membranes to the corresponding hydroxy-compounds. This reaction demands supply of reduced glutathione and will therefore cease post mortem when the cell is depleted of that substance. The membranes also contain the phenolic compound a-tocopherol (Vitamin E) which is considered the most important natural antioxidant.

Tocopherol can donate a hydrogen atom to the radicals L- or LOO- functioning as the molecule AH. It is generally assumed, that the resulting tocopheryl radical reacts with ascorbic acid

(Vitamin C) at the lipid/water interface regenerating the tocopherol molecule. Other compounds, for example the carotenoids, may also function as antioxidants. Wood smoke contains phenols which may penetrate the fish surface during smoking and thereby provide some protection against lipid oxidation.

Hydrolysis

During storage, a considerable amount of free fatty acids (FFA) appears. The phenomenon is more profound in ungutted than in gutted fish probably because of the involvement of digestive enzymes. Triglyceride in the depot fat is cleaved by triglyceride lipase originating from the digestive tract or excreted by certain microorganisms. Cellular lipases may also play a minor role.

In lean fish, for example Atlantic cod, production of free fatty acids also occurs, even at low temperatures. The enzymes responsible are believed to be cellular phospholipases - in particular phospholipase A_2 - although a correlation between activity of these enzymes and the rate of appearance of FFA has as yet not been firmly established.

The fatty acids bound to phospholipids at glycerol-carbon atom 2 are largely of the polyunsaturated type, and hydrolysis therefore often leads to increased oxidation as well. Furthermore, the fatty acids themselves may cause a "soapy" off-flavour.

CHEMICAL COMPOSITION

Principal Constituents

The chemical composition of fish varies greatly from one species and one individual to another depending on age, sex, environment and season.

The principal constituents of fish and mammals may be divided into the same categories, and examples of the variation between the constituents in fish are shown in Table below. The composition of beef muscle has been included for comparison.

Table: Principal constituents (percentage) of fish and beef muscle

Constituent	Fish (fillet)			Beef (isolated muscle)
	Min.	Normal variation	Max.	
Protein	6	16-21	28	20
Lipid	0.1	0.2-25	67	3
carboydrate		<0.5		1
Ash	0.4	1.2-1.5	105	1
Water	28	66-81	96	75

SOURCES: Stansby, 1962; Love, 1970

As can be seen from Table, a substantial normal variation is observed for the constituents of fish muscle. The minimum and maximum values listed are rather extreme and encountered more rarely. The variation in the chemical composition of fish is closely related to feed intake, migratory swimming and sexual changes in connection with spawning. Fish will have starvation periods for natural or physiological reasons (such as migration and spawning) or because of external factors such as shortage of food.

Usually spawning, whether occurring after long migrations or not, calls for higher levels of energy. Fish having energy depots in the form of lipids will rely on this. Species performing long migrations before they reach specific spawning grounds or rivers may utilize protein in addition to lipids for energy, thus depleting both the lipid and protein reserves, resulting in a general reduction of the biological condition of the fish. Most species, in addition, do usually not ingest much food during spawning migration and are therefore not able to supply energy through feeding.

During periods of heavy feeding, at first the protein content of the muscle tissue will increase to an extent depending upon how much it has been depleted, e.g., in relation to spawning migration. Then the lipid content will show a marked and rapid increase. After spawning the fish resumes feeding behaviour and often migrates to find suitable sources of food. Plankton-eating species such as herring will then naturally experience another seasonal variation than that caused by spawning, since plankton production depends on the season and various physical parameters in the oceans.

The lipid fraction is the component showing the greatest variation. Often, the variation within a certain species will display a characteristic seasonal curve with a minimum around the time of spawning.

Although the protein fraction is rather constant in most species, variations have been observed such as protein reduction occuring in salmon during long spawning migrations (Ando *et al.,* 1985 b; Ando and Hatano, 1986) and in Baltic cod during the spawning season, which for this species extends from January to June/July (Borresen, 1992).

Some tropical fish also show a marked seasonal I variation in chemical composition. West African shad *(Ethmalosa dorsalis)* shows a range in fat content of 2-7 % (wet weight) over the year with a maximum in July (Watts, 1957). Corvina *(Micropogon furnieri)* and pescada-foguete*(Marodon ancylodon)* captured off the Brazilian coast had a fat content range of 0.2-8.7 % and 0.1-5.4 % respectively (Ito and Watanabe, 1968). It has also been observed that the oil content of these species varies with size, larger fish containing about 1 % more oil than smaller ones. Watanabe (1971) examined freshwater fish from Zambia and found a variation from 0.1 to 5.0 % in oil content of four species including both pelagics and demersals.

A possible method for discriminating lean from fatty fish species is to term fish that store lipids only in the liver as lean, and fish storing lipids in fat cells distributed in other body tissues as fatty fish.

Typical lean species are the bottom-dwelling ground fish like cod, saithe and hake. Fatty species include the pelagics like herring, mackerel and sprat. Some species store lipids in limited parts of their body tissues only, or in lower quantities than typical fatty species, and are consequently termed semi- fatty species (e.g., barracuda, mullet and shark).

The lipid content of fillets from lean fish is low and stable whereas the lipid content in fillets from fatty species varies considerably. However, the variation in the percentage of fat is reflected in the percentage of water, since fat and water normally constitute around 80 % of the fillet. As a rule of thumb, this can be used to estimate the fat content from an analysis of the amount of water in the fillet. In fact, this principle is being utilized with success in a fat-analysing instrument called the Torry Fish Fat Meter, where it is the water content that is actually being measured (Kent et *al.*, 1992).

Whether a fish is lean or fatty the actual fat content has consequences for the technological characteristics postmortem. The changes taking place in fresh lean fish may be predicted from knowledge of biochemical reactions in the protein fraction, whereas in fatty species changes in the lipid fractions have to be included. The implication may be that the storage time is reduced due to lipid oxidation, or special precautions have to be taken to avoid this.

The carbohydrate content in fish muscle is very low, usually below 0.5 %. This is typical for striated muscle, where carbohydrate occurs in glycogen and as part of the chemical constituents of nucleotides. The latter is the Source of ribose liberated as a consequence of the autolytic changes *post mortem*.

As demonstrated above, the chemical composition of the different fish species will show variation depending on seasonal variation, migratory behaviour, sexual maturation, feeding cycles, etc. These factors are observed in wild, free-living fishes in the open sea and inland waters. Fish raised in aquaculture may also show variation in chemical composition, but in this case several factors are controlled, thus the chemical composition may be predicted. To a certain extent the fish farmer is able to design the composition of the fish by selecting the farming conditions. It has been reported that factors such as feed composition, environment, fish size, and genetic traits all have an impact on the composition and quality of the aquacultured fish (Reinitz et al., 1979).

The single factors having the most pronounced Impact on the chemical composition is considered to be the feed composition. The fish farmer is interested in making the fish grow as fast as possible on a minimum amount of feed, as the feed is the major cost component in aquaculture. The growth potential is highest when the fish is fed a diet with a high lipid content for energy purposes and a high amount of protein containing a well balanced composition of amino acids.

However, the basic metabolic pattern of the fish sets some limits as to how much lipid can be metabolized relative to protein. Because protein is a much more expensive feed ingredient than lipid, numerous experiments have been performed in order to substitute as much protein as possible with lipids.

Usually most fish species will use some of the protein for energy purposes regardless of the lipid content. When the lipid content exceeds the maximum that can be metabolized for energy purposes, the remainder will be deposited in the tissues, resulting in a fish with very high fat content. Apart from having a negative impact on the overall quality, it may also decrease the yield, as most surplus fat will be stored in depots in the belly cavity, thus being discarded as waste after evisceration and filleting.

A normal way of reducing the fat content of aquacultured fish before harvesting is to starve the fish for a period. It has been demonstrated for both fatty and lean fish species that this affects the lipid content.

It should be mentioned that in addition to allowing for the possibility of, within certain limits, predetermining the fish composition in aquaculture operations, keeping fish in captivity under controlled conditions also offers the possibility of conducting experiments in which variation in chemical composition observed in wild fish may be provoked. The experiments may be designed such that the mechanisms behind the variations observed in wild fish may be elucidated.

LIPIDS

The lipids present in teleost fish species may be divided into two major groups: the phospholipids and the triglycerides. The phospholipids make up the integral structure of the unit membranes in the cells; thus, they are often called structural lipids. The triglycerides are lipids used for storage of energy in fat depots, usually within special fat cells surrounded by a phospholipid membrane and a rather weak collagen network. The triglycerides are often termed depot fat. A few fish have wax esters as part of their depot fats.

The white muscle of a typical lean fish such as cod contains less than 1 % lipids. Of this, the phospholipids make up about 90 % *(Ackman, 1980)*. The phospholipid fraction in a lean fish muscle consists of about 69 % phosphatidyl-choline, 19 % *phosphatidyl-ethanolamine* and 5 % phosphatidyl-serine. In addition, there are several other phospholipids occurring in minor quantities.

The phospholipids are all contained in membrane structures, including the outer cell membrane, the endoplasmic reticulum and other intracellular tubule systems, as well as membranes of the organelles like mitochondria. In addition to phospholipids, the membranes also contain cholesterol, contributing to the membrane rigidity. In lean fish muscle cholesterol may be found in a quantity of about 6 % of the total lipids. This level is similar to that found in mammalian muscle.

As already explained, fish species may be categorized as lean or fatty depending on how they store lipids for energy. Lean fish use the liver as their energy depot, and the fatty species store lipids in fat cells througout the body. The fat cells making up the lipid depots in fatty species are typically located in the subcutaneous tissue, in the belly flap muscle and in the muscles moving the fins and tail. In some species which store extraordinarily high amounts of lipids the fat may also be deposited in the belly cavity. Depending on the amount of polyunsaturated fatty acids, most fish fats are more or less liquid at low temperature.

Finally, fat depots are also typically found spread throughout the muscle structure. The concentration of fat cells appears to be highest close to the myocommata and in the region between the light and dark muscle (Kiessling *et al.*, 1991). The dark muscle contains some triglycerides inside the muscle cells even in lean fish, as this muscle is able to metabolize lipids directly as energy. The corresponding light muscle cells are dependent on glycogen as a source of energy for the anaerobic metabolism. In dark muscle the energy reserves are completely catabolized to CO_2 and water, whereas in light muscle lactic acid is formed.

The mobilization of energy is much faster in light muscle than in dark muscle, but the formation of lactic acid creates fatigue, leaving the muscle unable to work for long periods at maximum speed.

Thus, the dark muscle is used for continuous swimming activities and the light muscle for quick bursts, such as when the fish is about to catch a prey or to escape a predator. An example of the seasonal variation in fat deposition in mackerel and capelin, where it is seen that the lipid content in the different tissues varies considerably. The lipid stores are typically used for long spawning migrations and when building up gonads (Ando *et al.*, 1985 a). When the lipids are mobilized for these purposes there are questions as to whether the different fatty acids present in the triglyceride are utilized selectively. This is apparently not the case in salmon, but in cod a selective utilization of $C_{22:6}$ has been observed (Takama *et al.*, 1985).

The phospholipids may also be mobilized to a certain extent during sustained migrations (Love, 1970), although this lipid fraction is considered to be conserved much more than the triglycerides.

In elasmobranchs, such as sharks, a significant quantity of the lipid is stored in the liver and may consist of fats like diacyl-alkyl-glyceryl esters or squalene. Some sharks may have liver oils with a minimum of 80 % of the lipid as unsaponifiable substance, mostly in the form of squalene (Buranudeen and Richards-Rajadurai, 1986).

Fish lipids differ from mammalian lipids. The main difference is that fish lipids include up to 40% of long-chain fatty acids (14-22 carbon atoms) which are highly unsaturated. Mammalian fat will rarely contain more than two double bonds per fatty acid molecule while the depot fats of fish contain several fatty acids with five or six double bonds (Stansby and Hall, 1967).

The percentage of polyunsaturated fatty acids with four, five or six double bonds is slightly lower in the polyunsaturated fatty acids of lipids from freshwater fish (approximately 70 %) than in the corresponding lipids from marine fish (approximately 88 %), (Stansby and Hall, 1967). However, the composition of the lipids is not completely fixed but can vary with the feed intake and season.

In human nutrition fatty acids such as linoleic and linolenic acid are regarded as essential since they cannot be synthesized by the organism. In marine fish, these fatty acids constitute only around 2 % of the total lipids, which is a small percentage compared with many vegetable oils. However, fish oils contain other polyunsaturated fatty acids which are "essential" to prevent skin diseases in the same way as linoleic and arachidonic acid. As members of the linolenic acid family (first double bond in the third position, w-3 counted from the terminal methyl group), they will also have neurological benefits in growing children. One of these fatty acids, eicosapentaenoic acid (C20:5 w 3), has recently attracted considerable attention because Danish scientists have found this acid high in the diet of a group of Greenland Eskimos virtually free from arteriosclerosis. Investigations in the United Kingdom and elsewhere have documented that eicosapen-taenoic acid in the blood is an extremely potent antithrombotic factor (Simopoulos et al., 1991).

PROTEINS

The proteins in fish muscle tissue can be divided into the following three groups:

1. Structural proteins (actin, myosin, tropormyosin and actomyosin), which constitute 70-80 % of the total protein content (compared with 40 % in mammals). These proteins are soluble in neutral salt solutions of fairly high ionic strength (^{3}0.5 M).
2. Sarcoplasmic proteins (myoalbumin, globulin and enzymes) which are soluble in neutral salt solutions of low ionic strength (<0.15 M). This fraction constitutes 25-30 % of the protein.

3. Connective tissue proteins (collagen), which constitute approximately 3 % of the -protein in teleostei and about 10 % in elasmobranchii (compared with 17 % in mammals).

The structural proteins make up the contractile apparatus responsible for the muscle movement as explained. The amino-acid composition is approximately the same as for the corresponding proteins in mammaliam muscle, although the physical properties may be slightly different. The isoelectric point (pI) is around pH 4.5-5.5. At the corresponding pH values the proteins have their lowest solublity.

The conformational structure of fish proteins is easily changed by changing the physical environment. Treatment with high salt concentrations or heat may lead to denaturation, after which the native protein structure has been irreversibly changed. When the proteins are denatured under controlled conditions their properties may be utilized for technological purposes. A good example is the production of surimi-based products, in which the gel forming ability of the myofibrillar proteins is used. After salt and stabilizers are added to a washed, minced preparation of muscle proteins, and after a controlled heating and cooling procedure the proteins form a very strong gel (Suzuki, 1981).

The majority of the sarcoplasmic proteins are enzymes participating in the cell metabolism, such as the anaerobic energy conversion from glycogen to ATP. If the organelles within the muscle cells are broken, this protein fraction may also contain the metabolic enzymes localized inside the endoplasmatic reticulum, mitochondria and lysosomes.

The fact that the composition of the sarcoplasmic protein fraction changes when the organelles are broken was suggested as a method for differentiating fresh from frozen fish, under the assumption that the organelles were intact until freezing (Rehbein *et al.*, 1978, Rehbein, 1979, Salfi *et al.*, 1985). However, it was later stated that these methods should be used with great caution, as some of the enzymes are liberated from the organelles also during iced storage of fish (Rehbein, 1992).

The proteins in the sarcoplasmic fraction are excellently suited to distinguishing between different fish species, as all the different species have their characteristic band pattern when separated by the isoelectric focusing method. The method was succesfully introduced by Lundstrom (1980) and has been used by many laboratories and for many fish species. A review of the literature is given by Rehbein (1990).

The chemical and physical properties of collagen proteins are different in tissues such as skin, swim bladder and the myocommata in muscle (Mohr, 1971). In general, collagen fibrils form a delicate network structure with varying complexity in the different connective tissues in a pattern similar to that found in mammals.

However, the collagen in fish is much more thermolabile and contains fewer but more labile cross-links than collagen from warm-blooded vertebrates. The hydroxyprolin content is in general lower in fish than in mammals, although a total variation between 4.7 and 10 % of the collagen has been observed (Sato *et at,* 1989).

Different fish species contain varying amounts of collagen in the body tissues. This has led to a theory that the distribution of collagen may reflect the swimming behaviour of the species (Yoshinaka *et at, 1988).* Further, the varying amounts and varying types of collagen in different fishes may also have an influence on the textural properties of fish muscle (Montero and Borderias, 1989). Borresen (1976) developed a method for isolation of the collagenous network surrounding each individual muscle cell. The structure and composition of these structures has been further characterized in cod by Almaas (1982).

The role of collagen in fish was reviewed by Sikorsky *et al.* (1984). An excellent, more recent review is given by Bremner (1992), in which the most recent literature of the different types of collagen found in fish is presented.

Fish proteins contain all the essential amino-acids and, like milk, eggs and mammalian meat proteins, have a very high biological value.

Table : Essential amino-acids (percentage) in various proteins

Amino-acid	Fish	Milk	Beef	Eggs
Lysine	8.8	8.1	9.3	6.8
Tryptophan	1.0	1.6	1.1	1.9
Histidine	2.0	2.6	3.8	2.2
Phenylalanine	3.9	5.3	4.5	5.4
Leucine	8.4	10.2	8.2	8.4
Isoleucine	6.0	7.2	5.2	7.1
Threonine	4.6	4.4	4.2	5.5
Methionine-cystine	4.0	4.3	2.9	3.3
Valine	6.0	7.6	5.0	8.1

SOURCES: Braekkan, 1976; Moustgard, 1957

Cereal grains are ususally low in lysine and/or the sulphur-containing amino-acids (methionine and cysteine), whereas fish protein is an excellent source of these aminoacids. In diets based mainly on cereals, a supplement of fish can, therefore, raise the biological value significantly.

In addition to the fish proteins already mentioned there is a renewed interest in specific protein fractions that may be recovered from by-products, particularly in the viscera. One such example is the basic protein or protamines found in the milt of the male fish. The molecular weight is usually below 10 000 kD and the pI is higher than 10. This is a result of the extreme amino-acid composition that may show as much as 65 % arginine.

The presence of the basic proteins has long been known, and it is also known that they are not present in all fish species (Kossel, 1928). The best sources are salmonids and herring, whereas ground fish like cod are not found to contain protamines.

The extreme basic character of protamines makes them interesting for several reasons. They will adhere to most other proteins less basic. Thus they have the effect of enhancing functional properties of other food proteins (Poole *et al.*, 1987; Phillips *et al.*, 1989). However, there is a problem in removing all lipids present in the milt from the protein preparation, as this results in an off-flavour in the concentrations to be used in foods. Another interesting feature of the basic proteins is their ability to prevent growth of microorganisms (Braekkan and Boge, 1964; Kamal *et al.*, 1986). This appears to be the most promising use of these basic proteins in the future.

N-CONTAINING EXTRACTIVES

The N-containing extractives can be defined as the water-soluble, low molecular weight, nitrogen- containing compounds of non-protein nature. This NPN-fraction (non-protein nitrogen) constitutes from 9 to 18 % of the total nitrogen in teleosts. The major components in this fraction are: volatile bases such as ammonia and trimethylamine oxide (TMAO), creatine, free amino-acids, nucleotides and purine bases, and, in the case of cartilaginous fish, urea.

An example of the distribution of the different compounds in the NPN-fraction in freshwater and marine fish. It should be noted that the composition varies not only from species to species, but also within the species depending on size, season, muscle sample, etc.

TMAO constitutes a characteristic and important part of the NPN-fraction in marine species and deserves further mention. This component is found in all marine fish species in quantities from 1 to 5 % of the muscle tissue (dry weight) but is virtually absent from freshwater species and from terrestrial organisms (Anderson and Fellers, 1952; Hebard *et al.*, 1982).

One exception was recently found in a study of Nile perch and tilapia from Lake Victoria, where as much as 150-200 mg TMAO/100 g of fresh fish was found (Gram *et al.*, 1989).

Although much work has been conducted on the origin and role of TMAO, there is still much to be clarified. Stroem *et al.* (1979) have shown that TMAO is formed by biosynthesis in certain zooplankton species. These organisms possess an enzyme (TMA mono-oxygenase) which oxidizes TMA to TMAO. TMA is commonly found in marine plants as are many other methylated amines (monomethylamine and dimethylamine).

Plankton-eating fish may obtain their TMAO from feeding on these zooplankton (exogenous origin). Belinski (1964) and Agustsson and Stroem (1981) have shown that certain fish species are able to synthesize TMAO from TMA, but this synthesis is regarded as being of minor importance.

The TMA-oxidase system is found in the microsomes of the cells and is dependent on the presence of Nicotinamide ademine denucleotide phosphate (NADPH):

$$(CH_3)_3N + NADPH + H^+ + O_2 \rightarrow (CH_3)_3NO + NADP^+ + H_2O$$

It is puzzling that this mono-oxygenase can be widely found in mammals (where it is thought to function as a detoxifier), while most fish have low or no detectable actitity of this enzyme.

Japanese research (Kawabata, 1953) indicates that there is a TMAO-reducing system present in the dark muscle of certain pelagic fishes. The amount of TMAO in the muscle tissue depends on the species, season, fishing ground, etc. In general, the highest amount is found in elasmobranchs and quid (75-250 mg N/100 g); cod have somewhat less (60-120 mg N/100 g) while flatfish and pelagic fish have the least. An extensive compilation of data is given by Hebard *et* al. (1982). According to Tokunaga (1970), pelagic fish (sardines, tuna, mackerel) have their highest concentration of TMAO in the dark muscle while demersal, white-fleshed fish have a much higher content in the white muscle. In elasmobranchs, TMAO seems to play a role in osmoregulation, and it has been shown that a transfer of small rays to a mixture of fresh and sea water (1: 1) will result in a 50 % reduction of intracellular TMAO. The role of TMAO in teleosts is more uncertain. Several hypotheses for the role of TMAO have been proposed:

- TMAO is essentially a waste product, a detoxified form of TMA
- TMAO is an osmoregulator
- TMAO functions as an "anti-freeze"
- TMAO has no significant function. It is accumulated in the muscle when the fish is fed a TMAO-containing diet

According to Stroem (1984), it is now generally believed that TMAO has an osmoregulatory role. As the occurrence of TMAO had previously been found virtually only in marine species until the observation published by Gram *et al.* (1989), it was speculated that TMAO together with high amounts of taurine could have additional effects, at least in fresh water fish (Anthoni *et al.*, 1990 a). Quantitatively, the main component of the NPN-fraction is creatine. In resting fish, most of the creatine is phosphorylated and supplies energy for muscular contraction.

The NPN-fraction also contains a fair amount of free amino-acids. These constitute 630 mg/ 100 g light muscle in mackerel *(Scomber scombrus),* 350-420 mg/ 100 g in herring *(Clupea harengus)* and 310-370 mg/100 g in capelin *(Mallotus villosus).* The relative importance of the different amino- acids varies with species. Taurine, alanine, glycine and imidazole-containing amino-acids seem to dominate in most fish. Of the imidazole-containing amino-acids, histidine has attracted much attention because it can be decarboxylated microbiologically to histamine. Active, dark-fleshed species such as tuna and mackerel have a high content of histidine.

VITAMINS AND MINERALS

The amount of vitamins and minerals is species-specific and can furthermore vary with season. In general, fish meat is a good source of the B vitamins and, in the case of fatty species, also of the A and D vitamins. Some freshwater species such as carp have high thiaminase actitity so the thiamine content in these species is usually low. As for minerals, fish meat is regarded as a valuable source of calcium and phosphorus in particular but also of iron, copper and selenium. Saltwater fish have a high content of iodine.

The vitamin content is comparable to that of mammals except in the case of the A and D vitamins which are found in large amounts in the meat of fatty species and in abundance in the liver of species such as cod and halibut. It should be noted that the sodium content of fish meat is relatively low which makes it suitable for low-sodium diets.

Table: Some mineral constituents of fish muscle

Element	Average value (mg/100 g)	Range (mg/100 g)
Sodium	72	30 -134
Potassium	278	19 -502
Calcium	79	19 -881
Magnesium	38	4.5-452
Phosphorus	190	68-550

SOURCE: Murray and Burt, 1969

In aquacultured fish, the contents of vitamins and minerals are considered to reflect the composition of the corresponding components in the fish feed, although the observed data should be interpreted with great caution (Maage *et al.*, 1991). In order to protect the n-3 polyunsaturated fatty acids, considered of great importance both for fish and human health, vitamin E may be added to the fish feed as an antioxidant. It has been shown that the resulting level of vitamin E in the fish tissue corresponds to the concentration in the feed (Waagbo *et al.*, 1991).

BIOLOGICAL ASPECTS

Classification

Fish are generally defined as aquatic vertebrates that use gills to obtain oxygen from water and have fins with variable number of skeletal elements called fin rays (Thurman and Webber, 1984). Five vertebrate classes have species which could be called fish, but only two of these groups - the sharks and rays, and the bonyfish - are generally important and widely distributed in the aquatic environment.

Fish are the most numerous of the vertebrates, with at least 20000 known species, and more than half (58 %) are found in the marine environment. They are most common in the warm and temperate waters of the continental shelves (some 8 000 species). In the cold polar waters about 1 100 species are found. In the oceanic pelagic environment well away from the effect of land, there are only some 225 species. Surprisingly, in the deeper mesopelagic zone of the pelagic environment (between 100 and 1 000 m depth) the number of species increases. There are some 1 000 species of so-called mid- water fish (Thurman and Webber, 1984).

Classifying all these organisms into a system is not an easy task, but the taxonomist groups organisms into natural units that reflect evolutionary relationships. The smallest unit is the species. Each species is identified by a scientific name which has two parts the genus and the specific epithet (binominal nomenclature). The genus name is always capitalized and both are italicized. As an example, the scientific (species) name of the common dolphin is *Delphinus delphis*. The genus is a category that contains one or more species, while the next step in the hierarchy is the family which may contain one or more genus. Thus the total hierarchical system is: Kingdom: Phylum: Class: Order: Family: Genus: Species.

The use of common or local names often creates confusion since the same species may have different names in different regions or, conversely, the same name is ascribed to several different species, sometimes with different technological properties.

As a point of reference the scientific name should, therefore, be given in any kind of publication or report the first time a particular species is referred to by its common name. For further information see the International Council for the Exploration of the Sea "List of names of Fish and Shellfish" (ICES, 1966); the "Multilingual Dictionary of Fish and Fish Products" prepared by the Organisation for Economic Cooperation and Development (OECD, 1990) and the "Multilingual Illustrated Dictionary of Aquatic Animals and Plants" (Commission of the European Communities, 1993). The classification of fish into cartilaginous and bony (the jawless fish are of minor importance) is important from a practical viewpoint, since these groups of fish spoil differently and vary with regard to chemical composition.

Furthermore, fish can be divided into fatty and lean species, but this type of classification is based on biological and technological characteristics.

ANATOMY AND PHYSIOLOGY

The Skeleton

Being vertebrates, fish have a vertebral column - the backbone - and a cranium covering the brain. The backbone runs from the head to the tail fin and is composed of segments (vertebrae). These vertebrae are extended dorsally to form neural spines, and in the trunk region they have lateral processes that bear ribs. The ribs are cartilaginous or bony structures in the connective tissue (myocommata) between the muscle segments (myotomes). Usually, there is also a corresponding number of false ribs or "pin bones" extending more or less horizontally into the muscle tissue. These bones cause a great deal of trouble when fish are being filleted or otherwise prepared for food.

Muscle Anatomy and Function

The anatomy of fish muscle is different from the anatomy of terrestrial mammals, in that the fish lacks the tendinous system connecting muscle bundles to the skeleton of the animal. Instead, fish has muscle cells running in parallel and connected to sheaths of connective tissue (myocommata), which are anchored to the skeleton and the skin. The bundles of parallel muscle cells are called myotomes.

All muscle cells extend the full length between two myocommata, and run parallel with the longitudinal direction of the fish. The muscle mass on each side of the fish makes up the fillet, of which the upper part is termed the dorsal muscle and the lower part the ventral muscle.

The fillet is heterogenous in that the length of the muscle cells vary from the head end (anterior) to the tail end (posterior). The longest muscle cells in cod are found at about the twelfth myotome counting from the head, with an average length around 10 mm in a fish that is 60 cm long (Love, 1970). The diameter of the cells also vary, being widest in the ventral part of the fillet. The myocommata run in an oblique, almost "plow-like" pattern perpendicular to the long axis of the fish, from the skin to the spine. This anatomy is ideally suited for the flexing muscle movements necessary for propelling the fish through the water.

As in mammals, the muscle tissue of fish is composed of striated muscle. The functional unit, i.e., the muscle cell, consists of sarcoplasma containing nuclei, glycogen grains, mitochondria, etc., and a number (up to 1 000) of myofibrils. The cell is surrounded by a sheath of connective tissue called the sarcolemma. The myofibrils contain the contractile proteins, actin and myosin. These proteins or filaments are arranged in a characteristic alternating system making the muscle appear striated upon microscopic examination. Most fish muscle tissue is white but, depending on the species, many fish will have a certain amount of dark tissue of a brown or reddish colour. The dark muscle is located just under the skin along the side of the body.

The proportion of dark to light muscle varies with the activity of the fish. In pelagic fish, i.e., species such as herring and mackerel which swim more or less continuously, up to 48 % of the body weight may consist of dark muscle (Love, *1970)*. In demersal fish, i.e., species which feed on the bottom and only move periodically, the amount of dark muscle is very small.

There are many differences in the chemical composition of the two muscle types, some of the more noteworthy being higher levels of lipids and myoglobin in the dark muscle.

From a technological point of view, the high lipid content of dark muscle is important because of problems with rancidity.

The reddish meat colour found in salmon and sea trout does not originate from myoglobin but is due to the red carotenoid, astaxanthin. The function of this pigment has not been clearly established, but it has been proposed that the carotenoid may play a role as an antioxidant. Further, the accumulation in the muscle may function as a depot for pigment needed at the time of spawning when the male develops a strong red colour in the skin and the female transport carotenoids into the eggs. The latter seems to depend heavily on the amount of carotenoids for proper development after fertilization. It is clearly seen that the muscle colour of salmonids fades at the time of spawning. The fish cannot synthesize astaxanthin and is thus dependent on ingestion of the pigment through the feed. Some salmonids live in waters where the natural prey does not contain much carotenoid, e.g., in the Baltic Sea, thus resulting in a muscle colour less red than salmonids from other waters. This may be taken as an indication that the proposed physiological function of astaxanthin in salmonids explained above may be less important.

In salmon aquaculture, astaxanthin is included in the feed, as the red colour of the flesh is one of the most important quality criteria for this species.

Muscle contraction starts when a nervous impulse sets off a release of Ca + + from the sarcoplasmic reticulum to the myofibrils. When the Ca + + concentration increases at the active enzyme site on the myosin filament, the enzyme ATP-ase is activated.

This ATP-ase splits the ATP found between the actin and myosin filaments, causing a release of energy. Most of this energy is used as contractile energy making the actin filaments slide in between the myosin filaments in a telescopic fashion, thereby contracting the muscle fibre. When the reaction is reversed (i.e., when the Ca + + is pumped back, the contractile ATP-ase activity stops and the filaments are allowed to slip passively past each other), the muscle is relaxed.

The energy source for ATP generation in the light muscle is glycogen, whereas the dark muscle may also use lipids. A major difference is, further, that the dark muscle contains much more mitochondria than light muscle, thus enabling the dark muscle to operate an extensive aerobic energy metabolism resulting in CO_2 and H_2O as the end products. The light muscle, mostly generating energy by the anaerobic metabolism, accumulates lactic acid which has to be transported to the liver for further metabolization. In addition, the dark muscle is reported to possess functions similar to those are found in the liver.

The different metabolic patterns found in the two muscle types makes the light muscle excellently fitted for strong, short muscle bursts, whereas the dark muscle is designed for continual, although not so strong muscle movements. Post mortem the biochemical and physiological regulatory functions operating in vivo ceases, and the energy resources in the muscle are depleted. When the level of ATP reaches its minimum, myosin and actin are interconnected irreversibly, resulting in rigor mortis.

The Cardiovascular System

The cardiovascular system is of considerable interest to the fish technologist since it is important in some species to bleed the fish (i.e., remove most of the blood) after capture.

The fish heart is constructed for single circulation. In bony fish it consists of two consecutive chambers pumping venous blood toward the gills via the ventral aorta.

Notes:

1. The heart pumps blood toward the gills.
2. The blood is aerated in the gills.
3. Arterial blood is dispersed into the capillaries where the transfer of oxygen and nutrients to the surrounding tissue takes place.
4. The nutrients from ingested food are absorbed from the intestines, then transported to the liver and later dispersed in the blood throughout the body.
5. In the kidneys the blood is "purified" and waste products are excreted via the urine.

After being aerated in the gills, the arterial blood is collected in the dorsal aorta running just beneath the vertebral column and from here it is dispersed into the different tissues via the capillaries. The venous blood returns to the heart, flowing in veins of increasingly larger size (the biggest is the dorsal vein which is also located beneath the vertebral column).

The veins all gather into one blood vessel before entering the heart. The total volume of the blood in fish ranges from 1.5 to 3.0 % of the body weight. Most of it is located in the internal organs while the muscular tissues, constituting two- thirds of the body weight, contain only 20 % of the blood volume. This distribution is not changed during exercise since the light muscle in particular is not very vascularized.

During blood circulation the blood pressure drops from around 30 mg Hg in the ventral aorta to 0 when entering the heart (Randall, 197Q). After the blood has passed through the gills, the blood pressure derived from the pumping activity of the heart is already greatly decreased. Muscle contractions are important in pumping the blood back to the heart and counterflow is prevented by a system of paired valves inside the veins. Clearly, the single circulation of fish is fundamentally different from the system in mammals, where the blood passes through the heart twice and is propelled out into the body under high pressure due to the contractions of the heart.

In fish, the heart does not play an important role in the transportation of blood from the capillaries back to the heart. This has been confirmed in an experiment where the impact of different bleeding procedures on the colour of cod fillets was examined. No difference could be found regardless of whether the fish had been bled by means of cutting the throat in front of or behind the heart before gutting, or had not been cut at all before slaughter.

In some fisheries, bleeding of the fish is very important as a uniform white fillet is desirable. In order to obtain this, a number of countries have recommended that fish are bled for a period (15-20 min) prior to being gutted. This means that throat cutting and gutting must be carried out in two separate operations and that special arrangements (bleeding tanks) must be provided on deck. This complicates the working process (two operations instead of one), time-consuming for the fishermen and increases the time-lag before the fish is chilled. Furthermore it requires extra space on an otherwise crowded working deck.

Several researchers have questioned the necessity of handling the fish in a two-step procedure involving a special bleeding period (Botta *et al.*, 1986; Huss and Asenjo, 1977 a; Valdimarsson *et al.* 1984). There seems to be general agreement about the following:

- bleeding is more affected by time onboard prior to bleeding/gutting than by the actual bleeding/gutting procedure.
- best bleeding is obtained if live fish are handled, but it is of major importance to cut the fish before it enters rigor mortis since it is the muscle contractions that force the blood out of the tissues.

Disagreement exists as to the cutting method. Huss and Asenjo (1977 a) found best bleeding if a deep throat cut including the dorsal aorta was applied, but this was not confirmed in the work of Botta *et al.* (1986). The latter also recommended to include a bleeding period (two-step procedure) when live fish were handled (fishing with pound net, trap, seine, longline or jigging), while Valdimarsson *et al.* (1984) found that the quality of dead cod (4 h after being brought onboard) was slightly improved using the two-step procedure. However, it should be pointed out that the effect of bleeding should also be weighted against the advantages of having a fast and effective handling procedure resulting in rapid chilling of the catch.

Discoloration of the fillet may also be a result of rough handling during catch and catch handling while the fish is still alive. Physical mishandling in the net (long trawling time, very large catches) or on the deck (fishermen stepping on the fish or throwing boxes, containers and other items on top of the fish) may cause bruises, rupture of blood vessels and blood oozing into the muscle tissue (haematoma). Heavy pressure on dead fish, when the blood is clotted (e.g., overloading of fish boxes) does not cause discoloration, but the fish may suffer a serious weight loss.

Other Organs

Among the other organs, only the roe and liver play a major role as foodstuffs. Their size depends on the fish species and varies with life cycle, feed intake and season. In cod the weight of the roe varies from a few percent up to 27 % of the body weight and the weight of the liver ranges from 1 to 4.5 %. Likewise, the composition can change and the oil content of the liver vary from 15 to 75 %, with the highest values being found during autumn (Jangaard *et al.,* 1967).

GROWTH AND REPRODUCTION

During growth it is the size of each muscle cell that increases rather than the number of muscle cells. Also, the proportion of connective tissue increases with age. Most fish become sexually mature when they reach a size characteristic of the species and is this not necessarily directly correlated with age. In general, this critical size is reached earlier in males than in females. As the growth rate decreases after the fish has reached maturity, it is therefore often an economic advantage to rear female fish in aquaculture. Every year mature fish use energy to build up the gonads (the roe and milk). This gonadal development causes a depletion of the protein and lipid reserves of the fish since it takes place during a period of low or no food intake.

In North Sea cod it was found that prior to spawning the water content of the muscle increases and the protein content decreases. In extreme cases the water content of very large cod can attain 87 % of the body weight prior to spawning (Love, 1970). The length of the spawning season varies greatly between species. Most species have a marked seasonal periodicity, while some have ripe ovaries for nearly the whole year. The depletion of the reserves of the fish during gonadal development can be extremely severe, especially if reproduction is combined with migration to the breeding grounds. Some species, e.g., Pacific salmon *(Oncorhynchus* spp.), eel *(Anguilla anguilla)* and others, manage to migrate only once, after which they degenerate and die. This is partly because these species do not eat during migration so that, in the case of a salmon, it can lose up to 92 % of its lipid, 72 % of its protein and 63 % of its ash content during migration and reproduction (Love, 1970).

On the other hand, other fish species are capable of reconstituting themselves completely after spawning for several years. The North Sea cod lives for about eight years before spawning causes its death, and other species can live even longer (Cushing, 1975). In former times, 25-year-old herring

(Clupea harengus) were not unusual in the Norwegian Sea, and plaice *(Pleuronectes platessa) up* to 35 years old have been found. One of the oldest fish reported was a sturgeon *(Acipenser sturio)* from Lake Winnebago in Wisconsin. According to the number of rings in the otolith, it was over 100 years old.

AQUATIC RESOURCES AND THEIR UTILIZATION

More than two-thirds of the world's surface is covered by water and the total yearly production of organic material in the aquatic environment has been estimated at about 40 000 million t (Moeller Christensen, 1968). Tiny microscopic plants, the phytoplankton, are the primary producers of organic material using the energy supplied by the sun.

This enormous primary production is the first link in the food chain and forms the basis for all life in the sea. How much harvestable fish results from this primary production has been the subject of much speculation. However, there are great difficulties in estimating the ecological efficiency, i.e., the ratio of total production at each successive trophic level. Gulland (1971) reports a range from 10 to 25 % but suggests 25 % as the absolute upper limit of ecological efficiency; for example, not all of the production at one trophic level is consumed by the next. Ecological efficiency also varies between levels, being higher at the lower levels of the food chain with smaller organisms using proportionally more of their food intake for growth rather than for maintenance. Diseases, mortality, pollution, etc. may also influence ecological efficiency. Since production is greater in the early stages of the food chain, the potential catch is also greater if harvesting is carried out at these stages. Up to 1970, the world catch of marine fish continued to rise at an overall rate of 6 percent per year, according to FAO statistics. Great optimism was expressed by various authors who estimated the potential world catch to be somewhere between 200 million t/year to 2 thousand t/year (Gulland, 1971); most of this wide variation being due to uncertainties concerning the trophic level at which the harvest would be taken.

The yearly increase in catches has slowed down since 1970, and the total catch reached a peak of 100 million t in

1989. Since then it has started to drop as a number of fish stocks have begun to collapse, in many cases due to overfishing. However, a slight upward trend is noticed for 1992 and for 1993 world catch is estimated to reach 101 million t. While total catch has started to decline since the peak in 1989, the catch from developing countries as a group is still increasing and since 1985 has exceeded that from developed countries.

Thus in 1992 little more than 60 % of the total world catch was taken by developing countries, and it is estimated that this figure will increase to 66% in 1993. This also means that an increasing part of the world fish catch is taken from warm tropical waters.

Are we then reaching the limits of production from "wild" aquatic resources now or do the optimistic predictions from the 1970s still hold? The answer to this question is not only in the affirmative, but for many resources the limit was reached decades earlier than the peak in global landings (FAO, 1993a).

A combination of factors has helped to mark the depletion of many conventional resources. One of these is that continued investments in fishing fleets throughout the world has meant that although catch rates and abundance of high value fish species have often declined, the overall level of fishing effort has increased so that roughly similar levels of landings are being taken at much greater cost to many fishing nations.

The real problems with decreasing fish stocks are familiar. First there is "the tragedy of the commons" - whatever lacks a known owner, whether buffalo or fish - which everyone will race to exploit and ultimately destroy.

The next problem which can be identified is the exceptionally poor management of the aquatic resources. What has been done has been too late and too little.

The 1982 Law of the Sea, which extended the territorial seas from 12 to 200 miles, gave the coastal States an opportunity to take a protective interest in their fishing grounds. Instead, many of them rushed to plunder the resources by offering generous subsidies and tax relief for new vessels. Also, the much used quota-system is subject to severe criticism. Often, the net result is increased fishing and increased waste, as perfectly good fish are thrown overboard if quotas are already reached. Many fish stocks (such as pollack, haddock and halibut off New England) are now considered "commercially extinct"; that is, there are now too few fish to warrant catching.

From an initial stage of under-utilization the fishing passes through a phase of rapid expansion until the limit of the resource is reached. This is then followed by a period o overfishing with high fishing effort, but reduced catches until finally - and hopefully - a phase of proper management is reached.

Details on resource management are beyond the scope of this book, but should include the concept of sustainability, environmental aspects and responsible fishing. However, in an FAO publication (FAO, 1994) it is stated that change from a focus on short-term development of fishing fleets to proper management is a necessary, but insufficient condition for sustainable development.

In the same report it is further stated that "Sustainable Development" as promoted at the United Nations Conference on Environment and Development (UNCED) in 1992 cannot be achieved under open-access regimes, whether these are within or outside national territorial waters.

In contrast, the world aquaculture production inclusive of aquatic plants has steadily increased over the last decade totalling 19.3 million ton in 1992, almost half of this (49% is produced in marine aquaculture, 44% in inland aquaculture, and the rest in brackish environment. About 49% of world aquaculture production are fish.

Production of aquati plants is increasing rapidly and reached 5.4 million t in 1992, while smaller increases if production of molluscs and crustaceans are seen. The total value of the aquaculture production is estimated to more than $US 32.5 billion in 1992. To summarize, it can be said that further increases in supply of fish can be expected from better utilization/ reduction of losses and further expansion of aquaculture.

Table shows the breakdown of world fish production.

Table : Breakdown of world fish production (percentage of world total in live weight) (FAO, 1993 a)

Year	For human consumption					Other purposes
	Total	Fresh	Freezing	Curing	Canning	Animal Feed
1982	71.1	19.4	25.3	12.8	13.6	28.9
1992	72.8	27.0	24.1	9.3	12.4	27.2

Table shows relatively modest differences in the breakdown of the fish production during the decade 1982-92. However, there was a significant increase in fresh fish consumption. Total fish for human consumption increased by 1.2% while fish used for curing and canning continued to decrease. In value terms, fishery exports reached an estimated $US 40.1 billion in 1993 (FISHDAB, 1994). Exports of fish and fishery products from developing countries continued to increase reaching a total value of $US 19.4 billion in 1993. In the same year exports from developed countries dropped by 5% to an estimated total value of $US 20.7 billion. Developing countries recorded an increasingly positive trade balance in fish trade, which reached $US 12.7 billion in 1993 (FISHDAB, 1994).

It should be noted that Table does not give a true picture of the amount of fish available for human food. An enormous amount of fish is wasted due to discards on board or post-harvest losses during processing and distribution. It has been estimated that the global amount of discards is in the range of 17-39 million t/year with an average of 27 million t/year (Alverson et al., 1994). It has been further estimated that the total post-harvest losses in fish products are about 10 % (James, D., personal communication 1994). These high losses are mainly due to problems of fisheries management, and lack of proper technology and of economic incentives.

8

Trends in Last Decade

World catches of fish have increased in the 1970s and 1980s but seem to have stabilized since 1988 to just under 100 million t. As the human population is ever increasing, it means that less fish will be available *per caput* every year. Nevertheless, a large part of this valuable commodity is wasted: it has been estimated by FAO that post-harvest losses (discards at sea and losses due to deterioration) remain at a staggering 25 % of the total catch.

Better utilization of the aquatic resources should therefore aim primarily at reducing these enormous losses by improving the quality and preservation of fish and fish products and by upgrading discarded low value fish to food products. Very often, ignorance and lack of skill in fish handling or in the administration of fisheries are among the causes for lack of progress in this direction.

FAO has long recognized the need for training in fish technology, and since 1971 a series of training courses, financed by the Danish International Development Agency (DANIDA), has been conducted in the developing countries. In 1988 a training manual entitled "Fresh fish- quality and quality changes" was published. This book has been extensively used and is now out of print. This present book is a revised and updated version of the first publication. It still only deals with fresh fish, as it is felt that a solid background knowledge of the raw material is essential for further development in preservation of and adding value to the product. In the context of this book, fresh fish is either fish kept alive until it is consumed, or dead fish preserved only by cold water or ice.

The book describes fundamentals in fish biology, chemical composition of fish and *post mortem* changes, with a view to explaining the rationale for optimal catch handling procedures and obtaining maximum shelf life. The effect of various factors (temperature, atmosphere, etc.) on fresh fish quality is discussed as are the various sensory, chemical and micro-biological methods for assessing fish quality. Wherever possible, data on tropical fish have been included.

Fresh fish handling procedures encompass all the operations aimed at maintaining food safety and quality characteristics from the time fish is caught until it is consumed. In practice, it means reducing the spoilage rate as much as possible, preventing contamination with undesirable microorganisms, substances and foreign bodies and avoiding physical damage of edible parts.

The immediate effect of fish handling procedures (e.g., washing, gutting, chilling) on quality can easily be assessed by sensory methods. Fish quality, in terms of safety and keeping time, is highly influenced by non-visible factors such as autolysis and contamination and growth of microorganisms. These effects can only be assessed long after the damage has occurred, and the proper procedures must thus be based on knowledge about the effects of the many different factors involved. Large or small improvements are usually feasible when analysing current fish handling methods.

QUALITY CHANGES AND SHELF LIFE OF CHILLED FISH

The effect of storage temperature

Chill storage (0-25°C)

It is well known that both enzymatic and microbiological activity are greatly influenced by temperature. However, in the temperature range from 0 to 25°C, microbiological activity is relatively more important, and temperature changes have greater impact on microbiological growth than on enzymatic activity.

Many bacteria are unable to grow at temperatures below 10°C and even psychrotrophic organisms grow very slowly, and sometimes with extended lag phases, when temperatures approach 0°C Figure shows the effect of temperature on the growth rate of the fish spoilage bacterium *Shewanella Putrefaciens*. At 0°C the growth rate is less than one-tenth of the rate at the optimum growth temperature.

Microbial activity is responsible for spoilage of most fresh fish products. The shelf life of fish products, therefore, is markedly extended when products are stored at low temperatures. In industrialized countries it is common practice to store fresh fish in ice (at 0°C) and the shelf life at different storage temperatures (at t°C) has been expressed by the relative rate of spoilage (RRS).

While broad differences are observed in shelf lives of the various seafood products, the effect of temperature on RRS is similar for fresh fish in general. Table shows an example with different seafood products.

Table : Shelf lives in days and relative rates of spoilage (RRS) of seafood products stored at different temperatures

	0°C		5°C		10°C	
	shelf life	RRS	shelf life	RRS	shelf life	RRS
Crab claw[a]	10.1	1	5.5	1.8	2.6	3.9
Salmon[b]	11.8	1	8.0	1.5	3.0	3.9
Sea bream[c]	32.0	1	-	-	8.0	4.0
Packed cod[d)]	14	1	6.0	2.3	3.0	4.7

a) Cann et al. (1985); b) Cann et al. (1984); c) Olley and Quarmby (1981); d) Cann et al. (1983)

The relationship between shelf life and temperature has been thoroughly studied by Australian researchers (Olley and Ratkowsky, 1973 a, 1973 b). Based on data from the literature they found that the relationship between temperature and RRS could be expressed as an S-shaped general spoilage curve. Particularly at low temperatures (e.g., < 10°C this curve is similar to, and confirms the results of Spencer and Baines (1964). These authors, 10 years earlier, found a straight line relationship between RRS and the storage temperatures of cod from the North Sea.

The effect of temperature on the rate of chemical reactions is often described by the Arrhenius Equation. This Equation, however, has been shown not to be accurate when used for the effect of a wide range of temperatures, on growth of microorganisms and on spoilage of foods (Olley and Ratkowsky, 1973 b; Ratkowsky *et* al., 1982). Ratkowsky *et al.* (1982) suggested the 2-parameter square root model (Equation 6.b) for the effect of sub-optimal temperature on growth of microorganisms

$$\sqrt{\mu_{max}} = b(T - T_{min}) \qquad 6b$$

T is the absolute temperature (Kelvin) and T_{min} in a parameter expressing the theoretical minimum temperature of growth. The square root of the microbial growth rates plotted against the temperature form a straight line from which T_{min} is determined. Several psychrotrophic bacteria isolated from fish products have T_{min} values of about 263 Kelvin (-10°C) (Ratkowsky *et al.*, 1982; Ratkowsky *et al.*, 1983). Based on this T_{min} value, a spoilage model has been developed. It has been assumed that the relative microbial growth rate would be similar to the relative rate of spoilage.

The relative rate concept was then combined with the simple square root model to give a temperature spoilage model. As just described, this model was derived from growth of psychrotrophic model has been shown to give good estimates of the effect of temperature on bacteria (T_{min} = –10°C) but the RRS of chilled fresh fish as shown in Figure studies (Storey, 1985; Gibson, 1985).

$$\sqrt{\text{Relative rate or spoilage}} = \frac{b(t°C - (-10°C))}{b(0°C - (-10°C))} = 0.1 + t°C + 1$$

If the shelf life of a fish product is known at a given temperature, the shelf life at other storage temperatures can be calculated from the spoilage models. The effect of temperature, as calculated from Equation 6.c for products with different shelf lives when stored at 0°C. The effect of time/temperature storage conditions on product shelf life has been shown to be cumulative (Charm et al, 1972). This allows spoilage models to be used for prediction of the effect of variable temperatures on product keepability. An electronic time/temperature function integrator for shelf life prediction was developed. The instrument predicts RRS accurately, but a high price has limited its practical application (Owen and Nesbitt, 1984; Storey, 1985).

Table : Predicted shelf lives of fish products stored at different temperatures

Shelf life in days of product stored in ice (0°C)	Shelf life at chill temperatures (days)		
	5°C	10°C	15°C
6	2.7	1.5	1
10	4.4	2.5	1.6
14	6.2	3.5	2.2
18	8	4.5	2.9

The temperature history of a product, e.g., through a distribution system, can be determined by a temperature logger. Using a spoilage model and simple PC software, the effect of a given storage temperature profile can then be predicted. McMeekin *et al.* (1993) reviewed the literature on application of temperature loggers and on predictive temperature models. A product temperature profile also allows growth of pathogenic microorganisms to be estimated from safety models. Computers and temperature loggers are today available at reasonable prices and it is most likely that spoilage and safety models will be used frequently in the future.

The microflora responsible for spoilage of fresh fish changes with changes in storage temperature. At low temperatures (0-5°C), *Shewanella putrefaciens, Photobacterium phosphoreum, Aeromonas spp.* and *Pseudomonas* spp. cause spoilage. However, at high storage temperatures (15-30°C) different species of Vibrionaceae, Enterobacteriaceae and Gram-positive organisms are responsible for spoilage (Gram *et al.*, 1987; Gram *et al.*,

1990; Liston, 1992). Equation does not take into account the change in spoilage microflora. Nevertheless, reasonable estimates of RRS are obtained for whole fresh fish, for packed fresh fish and for superchilled fresh fish products. For tropical fish, however, the average relative rate of spoilage of a large number of species stored at 20°-30°C was approximately 25 times higher than at 0°C. Tropical fish are likely to be exposed to high temperatures and a new tropical spoilage model, covering the range of temperatures from 0°-30°C, was recently developed.

$$Ln \text{ (relative rate of spoilage for tropical fish)} = 0.12 * t°C$$

Temperature models based on the relative rate concept do not take into account the initial product quality. Inaccurate shelf life predictions, therefore, may be obtained for products with variable initial quality. Spencer and Baines (1964), however, suggested that both the effect of the initial product quality and the effect of storage temperature could be predicted. At a constant storage temperature measurements of quality will change linearily from an initial to a final level reached when the product is no longer acceptable. Shelf life at a given temperature and a given initial quality is determined and then the shelf life at other temperatures can be determined from a temperature spoilage model.

$$\text{Shelf life} = \frac{\text{final} - \text{initial level of a quality indicator}}{\text{rate of spoilage at the actual storage conditions}}$$

Much later, the demerit point system, also known as the quality index method, was developed and has proved most useful for obtaining a straight line relationship between quality scores and storage time. Bremner et al. (1987) suggested that the rate of change in quality scores, determined by the demerit point system, couldbe quantitatively described at different temperatures by Equation. Gibson (1985) related microbiological conductance detection times (DT), determined with the Malthus Growth Analyzer, to shelf life of cod. At storage temperatures from 0° to 10°C the daily rate of change in DT values was well predicted by Equation, and shelf lives were predicted at different temperatures from initial and final DT values and from the temperature spoilage model.

Many aspects of fresh fish spoilage remain to be studied; e.g., the activity of the microorganisms responsible for spoilage at different storage temperatures. Despite this lack of understanding, the relative rate concept has made it possible to quantify and mathematically describe the effect of temperature on the rate of spoilage of various types of fish products. These temperature spoilage models allow time/temperature function integration to be used for evaluation of production, distribution and storage conditions, and when combined with methods for determination of initial product quality, shelf life of various fish products can be predicted.

Apart from the actual storage temperature, the delay before chilling is of great importance. Thus, it can be observed that if white-fleshed, lean fish enter rigor mortis at temperatures above + 17°C, the muscle tissue may be ruptured through severe muscle contractions and weakening of the connective tissue (Love, 1973). The flakes in the fillets separate from each other and this "gaping" ruins the appearance. The fish also become difficult to fillet and the water- binding capacity decreases.

Table : Fillet yield of gutted cod (Hansen, 1981)

	Percentage fillet yield	
	Iced 1 hour after catch	**Iced 6 ½ hour after catch**
Yield of fillets	48.4	46.5
Yield after trimming	43.3	40.4

Rapid chilling is also crucial for the quality of fatty fish. Several experiments have shown that herring and garfish *(Belone belone)* have a significantly reduced storage life if they are exposed to sun and wind for 4-6 hours before chilling. The reason for the observed rapid quality loss is oxidation of the lipids, resulting in rancid off-flavours. It should be noted, however, that high temperatures are only partly responsible for the speed of the oxidation processes. Direct sunlight combined with wind may have been more important in this experiment as it is difficult to stop autocatalytic oxidation processes once they have been initiated.

Superchilling (0°C to -4°C)

Storage of fish at temperatures between 0°C and -4°C is called superchilling or partial freezing. The shelf life of various fish and shellfish can be extended by storage at subzero temperatures. The square root spoilage model gives a reasonable description of RRS of superchilled products. The shelf life predicted by the square root model at -1°C, -2°C and -3°C for a product that keeps 14 days in ice is 17, 22 and 29 days, respectively.

Superchilling extends the shelf life of fish products. The technique can be used, for example where productive fishing grounds are so far from ports and consumers that normal icing is insufficient for good quality products to be landed and sold. The application of superchilling to replace transport of live fish has also been studied in Japan (Aleman *et al.*, 1982).

The technology needed to use superchilling at sea as well as for storage on-land is available today. The "Frigido-system", developed in Portugal in the 1960s, uses heat exchanges in the fish holds. Sub- zero temperatures were kept constant (±0.5°C) and the fish:ice ratio was reduced from the normal 1:1 to 3:1. Sub-zero storage temperatures in fishing vessels can also be obtained in refrigerated sea water (RSW) where the freezing point of water is reduced by NaCl or other freezing point depressors.

Compared to ice storage, the RSW systems chill fish more rapidly, reduce the exposure to oxygen, reduce the pressures that often occur when fish are iced and also give significant labour-saving (Nelson and Barnett, 1973). Promising results have been obtained with superchilling, but both technical problems and problems in relation to product quality have been observed. Unloading of fish is difficult when heat exchanges are used in fishing vessels and RSW increases the corrosion of the vessels (Partmann, 1965; Barnett *et* al., 1971).

Also, superchilling extends product shelf life, but a negative effect on freshness/prime quality has been observed for some fish species. Merritt (1965) found that cod stored at -2°C for 10 days had an appearance and texture inferior to fish stored at 0°C in ice. The drip of the superchilled fish was increased and at -3°C the texture of whole cod made them unsuitable for filleting.

RSW storage gives several fish species a salty taste due to the take-up of sea water (Barnett *et al.,* 1971; Shaw and Botta, 1975; Reppond and Collins, 1983; Reppond *et al.,* 1985). This negative effect of RSW, however, has not been found in all studies (Lemon and Regier, 1977; Olsen *et al.,* 1993). As opposed to cod and several other fish species, the prime quality of superchilled shrimp from Pakistan was increased from 8 days in ice to 16 days in NaCl-ice at -3°C (Fatima *et al.,* 1988).

Also, both freshness (measured by a K-value of 20%) and shelf life of cultured carp *(Cyrinus carpio),* cultured rainbow trout *(Salmo gairdnerii)* and mackerel *(Scomber japonicus)* have been improved by superchilling at -3°C as compared to storage at 0°C (Uchiyama *et al.,* 1978 a, 1978 b; Aleman *et al.,* 1982).

The percentage of frozen water in superchilled fish is highly temperature-dependent (-1°C = 19%; -2°C = 55%; -3°C = 70%; -4°C = 76%) (Ronsivalli and Baker, 1981). It has been suggested that negative effects of superchilling on drip loss, appearance, and texture of cod and haddock are due to formation of large ice crystals, protein denaturation and increased enzymatic activity in the partially frozen fish (Love and Elerian, 1964). Simpson and Haard (1987), however, found only very little difference in biochemical and chemical deterioration of cod *(Gadus morhua)* stored at 0°C and at - 3°C In Japanese studies with seabass, carp, rainbow trout and mackerel, it has been shown that the drip loss as well as several biochemical and chemical deteriorative reactions were reduced in superchilled fish, compared to ice storage (Uchiyama and Kato, 1974; Kato *et al.,* 1974; Uchiyama *et al.,* 1978 a, 1978 b; Aleman *et al.,* 1982).

Superchilling has been used industrially with a few fish species such as tuna and salmon. The negative effects on sensory quality found for some other species may have limited the practical application of the technique. Nevertheless, it seems that shelf life of at least some seafood products is improved considerably by superchilling. Consequently, for selected products, superchilling may well be more suitable than other technologies.

THE EFFECT OF HYGIENE DURING HANDLING

Onboard Handling

Much emphasis has been placed on hygienic handling of the fish from the moment of catching in order to ensure good quality and long storage life. The importance of hygiene during handling onboard has been tested in a series of experiments where various hygienic measures were employed (Huss *et al.*, 1974). The quality and storage life of completely aseptically treated fish (aseptic handling) were compared with fish iced in clean plastic boxes with clean ice (clean handling) and with fish treated badly, i.e., iced in old, dirty wooden boxes (normal handling). As expected, a considerable difference is found in the bacterial contamination of the three batches. However, a similar difference in the organoleptic quality is not detected. During the first week of storage no difference whatsoever is found. Only during the second week does the initial contamination level become important and the heavily contaminated fish have a reduction in storage life of a few days compared with the other samples.

On the basis of these data it seems sensible to advocate reasonably hygienic handling procedures including use of clean fish boxes. Very strict hygienic measures do not seem to have great importance. In comparison with the impact of quick and effective chilling, the importance of hygiene is minor.

The above-named observations have influenced the discussion about the design of fish boxes. Normally, fish are iced in boxes stacked on top of each other. In this connection it has been argued that fish boxes should have a construction that prevents the ice melt-water from one box draining into the box underneath it. In a system like this, some bacterial contamination of fish in the bottom boxes would be avoided, as melt-water usually contains a large number of bacteria.

However, practical experience as well as experiments (Peters *et al.*, 1974) have shown that this type of contamination is unimportant, and it may be concluded that fish boxes allowing the drainage of melt-water from upper into lower boxes are advantageous because the chilling becomes more effective.

Inhibition or Reduction of the Naturally Occurring Microflora

In spite of the relatively minor importance of the naturally occurring microflora in the quality of the fish, much effort has been put into reduction or inhibition of this microflora. Many of these methods are only of academic interest. Among these are (at least until now) attempts to prolong the storage life by using radioactive irradiation. Doses of 100 000 - 200 000 rad are sufficient to reduce the number of bacteria and prolong storage life (Hansen, 1968; Connell, 1975), but the process is costly and, to many people, unacceptable in connection with human food. Another method which has been rejected because of concern about public health is treatment with antibiotics incorporated in the ice.

A method that has been used with some success over recent years is treatment with CO_2, which can be applied either in containers with chilled seawater or as part of a modified atmosphere during distribution or in retail packages.

It should also be mentioned that washing with chlorinated water has been tried as a means of decontaminating fish. However, the amount of chlorine necessary to prolong the storage life creates off- flavours in the fish meat (Huss, 1971). The newly-caught fish should be washed in clean seawater without any additives. The purpose of the washing is mainly to remove visible blood and dirt, and it does not cause any significant reduction in the number of bacteria and has no effect on storage life.

The Effect of Anaerobic Conditions and Carbon Dioxide

High CO_2 concentrations can reduce microbial growth and may therefore extend the shelf life of food products, where spoilage is caused by microbial activity (Killeffer, 1930; Coyne, 1933). Technological aspects of modified atmosphere packaging (MAP) have since been studied. Today, materials and techniques for storage of bulk or retail packed foods are available.

This section discusses the effect of anaerobic conditions and modified atmospheres on the shelf life of fish products. The safety aspects are reviewed in Farber (1991) and Reddy et al. (1992).

Effect on Microbial Spoilage

Vacuum packaging (VP) and MAP, with high CO_2 levels (25% - 100%), extends the shelf life of meat products by several weeks or months.

Table : Effect of packaging on the shelf life of chilled fish and meat products

Type of product	Storage temp.	Shelf life (weeks)		
		Air	VP[a]	MAP[b]
Meat (beef, pork, poultry)	1.0 - 4.4°C	1 - 3	1 - 12	3 - 21
Lean fish (cod, pollock, rockfish, trevally)	0.0 - 4.0°C	1 - 2	1 - 2	1 - 3
Fatty fish (herring, salmon, trout)	0.0 - 4.0°C	1 - 2	1 - 2	1 - 3
Shellfish (crabs, scampi, scallops)	0.0 - 4.0°C	½ - 2	-	½ - 3
Warmwater fish (sheepshead, swordfish, tilapia)	2.0 - 4.0°C	½ - 2	-	2 - 4

a) VP: Vacuum packed

b) MAP: Modified atmosphere packed (High CO_2 concentrations (25 - 100%)

In contrast, the shelf life of fresh fish is not affected by VP and only a small increase in shelf life can be obtained by MAP. Differences in spoilage microflora. and in pH are mainly responsible for the observed differences in the shelf life of fish and meat products. Spoilage of meat under aerobic conditions is caused by strict aerobic Gram-negative organisms, primarily *Pseudomonas spp*. These organisms are strongly inhibited by anaerobic conditions and by CO_2. Consequently, they do not play any role in the spoilage of packed meat. Instead the microflora, of VP and MAP meat products changes to be dominated by Gram-positive organisms (Lactic Acid Bacteria), which are much more resistant to CO_2 (Molin, 1983; Dainty and Mackey, 1992). Fish stored under aerobic conditions are also spoiled by Gram negative-organisms, primarily *Shewanella putrefaciens*.

The spoilage flora on some packed fish products was found to be dominated by Grampositive microorganisms and in this way the microflora, was similar to the flora on packed meats; see Stammen *et al.* (1990) for a review. For packed cod, however, the Gram-negative organism *Photobacterium phosphoreum* has been identified as the organism responsible for spoilage. The growth rate of this organism is increased under anaerobic conditions and this may explain the importance of the organism in VP cod.

In CO_2-packed fish, the growth of *Shewanella putrefaciens* and of many other microorganisms found on live fish is strongly inhibited. In contrast P. *phosphoreum* was shown to be highly resistent to CO_2.

It was also shown that the limited effect of CO_2 on growth of this bacteria correspond very well with the limited effect of CO_2 on the shelf life of packed fresh cod. P. *phosphoreum* reduces TMAO to TMA while very little H_2S is produced during growth in fish substrates.

Spoiled VP and MAP cod is characterized by high levels of TMA, but little or no development of the putrid or H_2S odours typical for some aerobically stored spoiled fish. The growth characteristics of P. *phosphoreum* and the metabolic activity of the organism thus explain both the short shelf life and the spoilage pattern of packed cod (Dalgaard, 1994 a).

The shelf life of VP and MAP cod is similar to various other sea food products. P. *phosphoreum* is widespread in the marine environment and it seems likely that this organism or other highly CO_2 resistent microorganisms are responsible for spoilage of packed sea food products (Baumann and Baumann, 1981;

van Spreekens, 1974; Dalgaard *et al.*, 1993). The best effect of MAP storage on shelf life has been obtained with fish from warm waters. The shelf life of these products, however, is still relatively short compared to meat products.

Very low bacterial level (10^5-10^6 cfu/g) has been found at the time of sensory rejection of some packed fish products. In these cases non-microbial reactions may have been responsible for spoilage.

Effect of Non-microbial Spoilage Reactions

CO_2 is dissolved in the water phase of the flesh of MAP fish and a decrease in pH of about 0.2- 0.3 units is observed, depending on the CO_2 concentration in the surrounding gaseous atmosphere. The water-holding capacity of muscle proteins is decreased by decreased pH and an increased drip loss is expected for fish stored in high CO_2 concentrations. Increased drip has been found for cod fillets, red hake, salmon, and shrimps (Fey and Regenstein, 1982; Layrisse and Matches, 1984; Dalgaard *et al.*, 1993) but not for herring, red snapper, trevally, Dungeness crab, and rockfish (Cann *et al.*, 1983; Gerdes *et al.* 1991; Parking and Brown, 1983 and Parkin *et al.*, 1981).

Coyne (1933) and many later studies have found the textural quality of fish stored in 100% CO_2 to be reduced. However, up to 60% CO_2 has no negative effect on the texture of cod. The colour of the belly flaps, of cornea, and of the skin may be altered for whole fish stored in high CO_2 concentrations (Haard, 1992). Packaging may also stimulate the formation of metmyoglobulin in red- fleshed fish and thereby result in a darkening of fish muscles. Although oxygen-containing modified atmospheres have been used, the development of rancid off-odours in fatty fish species has not been registered as a problem (Haard, 1992).

Carbon Dioxide used in Combination with Refrigerated Seawater Systems

Only the effect of addition of CO_2 to RSW will be considered in this section. Table shows the effect of RSW and RSW + CO_2 on the shelf life of various fish products, as compared to storage in ice. An evident shelf life-extending effect of CO_2 is only seen with some species. Several negative effects of adding CO_2 to RSW-systems have been observed. The fish colour and texture were negatively influenced, and CO_2 dissolved in the flesh made mackerel unsuitable for canning (Longard and Regier, 1974; Lemon and Regier, 1977).

CO_2 acidifies the seawater, and a lowered pH inhibits the enzymatic reactions that otherwise lead to black spots in shrimps and prawns. The shelf life of pink shrimps can be more than doubled by storage in RSW + CO_2, where, compared to ice storage, colour, texture, flavour, and odour were improved (Nelson and Barnett, 1973). RSW+CO_2 stored prawns, however, may be unacceptably tough and have a "soft shell" appearance (Ruello, 1974).

Future Application of Carbon Dioxide for Shelf Life Extension

For most MAP seafoods, the production of TMA is delayed by only a few days compared to aerobic or anaerobic storage. This indicates that fish products in general are contaminated with a highly CO_2 resistent microflora of TMAO reducing organisms. Very high CO_2 concentrations can inhibit microbial growth but high levels of CO_2 have a negative effects on other aspects of the fish quality. MAP has found little practical application with fish products as compared to meat products. The main reasons for this are probably that:

- MAP used with retail packs is an expensive technique
- the prime fish quality is not improved
- only small shelf life extensions are obtained
- MAP cannot replace good chilling or good hygienic production conditions
- toxin production of *Clostridium botulinum* is increased for bacteria growing under anaerobic conditions, and this may be of importance for the safety of packed fish (Huss *et al.*, 1980; Reddy *et al.*, 1992).

Packaging, however, can be used simply because packed products are more convenient to handle, e.g., in supermarkets.

According to the EEC Council Directive of *22 July* 1991 (91/493/EEC), VP and MAP fish products are considered as fresh products. Consequently, CO_2 can be used for preservation of fresh fish products, when a shelf life extension of only a few days is found to be sufficient.

The negative effect Of CO_2 on fish colour is primarily a problem for whole fish and the negative effect of CO_2 on texture and drip loss is only observed with high CO_2 concentrations. A pronounced effect on growth of *S. putrefaciens* and on many other bacteria is obtained with even moderate CO_2 concentrations (40-80%). It is therefore likely that, in the future, MAP will be used in combination with preservation techniques that has been developed specifically to inhibit growth of CO_2 resistent TMAO reducing marine spoilage bacteria such as P. *phosphoreum*.

The effect of MAP also seems to depend on fish species and further studies are needed to determine if MAP can give interesting shelf life extensions for other fish species, e.g., those from warm waters. Finally, high CO_2 concentrations could be used for fish intended for fishmeal as the negative effects of CO_2 on colour and texture in this case are less important.

The Effect of Gutting

It is a common experience that the quality and storage life of many fish decrease if they have not been gutted. During feeding periods the fish contain many bacteria in the digestive system and strong digestive enzymes are produced. The latter will be able to cause a violent autolysis *post mortem,* which may give rise to strong off-flavour especially in the belly area, or even cause belly-burst.

On the other hand, gutting means exposing the belly area and cut surfaces to the air thereby rendering them more susceptible to oxidation and discoloration. Thus, many factors such as the age of the fish, the species, amount of lipid, catching ground and method, etc., should be taken into consideration before deciding whether or not gutting is advantageous.

Fatty Species

In most cases,small- and medium-sized fatty fish such as herring, sardines and mackerel are not eviscerated immediately after catch. The reason for this is partly that a large number of small fish are caught at the same time and partly because of problems with discoloration and the acceleration of rancidity.

However, problems may arise with ungutted fish during periods of heavy feeding due to belly- burst. The reactions leading to belly-burst are complex and not fully understood. It is known that the strength of the connective tissue is decreased during these periods and that post mortem pH is normally lower in well-fed fish, this also weakens the connective tissue. Furthermore, it seems that the type of feed ingested may play an important role in the belly-burst phenomenon.

Lean Species

In most North European countries, the gutting of lean species is compulsory. It is based on the assumption that the quality of these species suffers if they are not gutted. In the case of cod, it has been shown that omission causes a considerable quality loss and a reduction in the storage life of five or six days. After only two days from catch, discoloration of the belly area is visible and the raw fillet acquires an offensive cabbagey odour.

These volatile, foul-smelling compounds are mostly found in the gut and surrounding area whereas the amount of volatile acids and bases is relatively low in the fillet itself. These chemical parameters are, therefore, not useful for distinguishing between gutted and ungutted fish (Huss and Asenjo, 1976).

In the case of haddock *(Melanogrammus aeglefinus)*, whiting *(Merlangius merlangus*, saithe *(Pollachius virens)* and blue whiting *(Micromesistius poutassou)*, it is observed that ungutted fish stored at 0°C suffer a quality loss compared with gutted fish, but the degree varies. Some off-odours and off-flavours are detected, but ungutted haddock, whiting and saithe are still acceptable as raw material for frozen fillets after nearly one week on ice (Huss and Asenjo, 1976). Quite different results are obtained with South American hake *(Merluccius gayi),* where no difference is observed between gutted and ungutted fish (Huss and Asenjo, 1977 b).

The Effect of Fish Species, Fishing Ground and Season

Influence of Handling, Size, pH, Skin Properties

The spoilage rate and shelf life of fish is affected by many parameters and, fish spoil at different rates. In general it can be stated that larger fish spoil more slowly than small fish, flat fish keep better than round fish, lean fish keep longer than fatty fish under aerobic storage and bony fish are edible longer than cartilaginous fish. Several factors probably contribute to these differences and whereas some are clear, many are still on the level of hypotheses.

Table : Intrinsic factors affecting spoilage rate of fish species stored in ice

Factors affecting spoilage rate	Relative spoilage rate	
	fast	slow
size	small fish	larger fish
post mortem pH	high pH	low pH
fat content	fatty species	lean species
skin properties	thin skin	thick skin

Rough handling will, result in a faster spoilage rate. This is due to the physical damage to the fish, resulting in easy access for enzymes and spoilage bacteria. The surface/volume ratio of larger fish is lower than that of smaller fish, and, as bacteria are found on the outside, this is probably the reason for the longer shelf life of the former. This is true within a species but may not be universally so.

Post mortem pH varies between species but is, higher than in warm- blooded animals. The long rigor period and the corresponding low pH (5.4-5.6) of the very large flatfish, halibut (*Hippoglossus hipoglossus*), has been offered as an explanation for its relatively long iced storage life. However, mackerel will often also experience a low pH and this seems to have little effect on shelf life.

The skin of the fatty pelagic fish is often very thin, and this may contribute to the faster spoilage rate. This allows enzymes and bacteria to penetrate more quickly. On the contrary, the thick skin of flatfish and the antibacterial compounds found in the slime of these fish may also contribute to the keepability of flatfish.

As described earlier, the slime of flat fish contains bacteriolytic enzymes, antibodies and various other antibacterial substances (Hjelmland et al., 1983; Murray and Fletcher, 1976). Although large differences exist in the content of TMAO, this does not seem to affect the shelf life of aerobically-stored fish but rather the chemical spoilage profile of the species.

Table : Shelf life of various fish species from temperate and tropical waters. Prepared from data published by Lima dos Santos (1981); Poulter *et al.*(1981); and Gram (1989)

Species	**Fish type**	**Shelf life (days in ice)**	
		temperate	**tropical**
Marine species		2-24	6-35
cod, haddock	lean	9-15	
whiting	lean	7-9	
hake	lean	7-15	
bream	lean/low fat		10-31
croaker	lean		8-22
snapper	lean		10-28
grouper	lean		6-28
catfish	lean		16-19
pandora	lean		8-21
jobfish	lean		16-35
spadefish	lean/low fat		21-26
batfish	lean		21-24
sole, plaice,	flat	7-21	21

flounder	flat	7-18	
halibut	flat	21-24	
mackerel[1)]	high/ low fat	4-19	14-18
summer herring	high fat	2-6	
winter herring	low fat	7-12	
sardine	high fat	3-8	9-16
Freshwater species		9-17	6-40
catfish	lean	12-13	15-27
trout	low fat	9-11	16-24
perch	lean/ low fat	8-17	13-32
tilapia	lean		10-27
mullet	lean		12-26
carp	lean/ low fat		16-21
lungfish	lean/ low fat		11-25
Haplochromis	lean		6
shad	medium fat		25
corvina	medium fat		30
bagré	medium fat		25
chincuna	fatty		40
pacu	fatty		40

1) fat content and shelf life subject to seasonal variation.

In general, the slower spoilage of some fish species has been attributed to a slower bacterial growth, and Liston (1980) stated that "different spoilage rates seem to be related at least partly to the rate of increase of bacteria on them".

Influence of Water Temperature on Iced shelf Life

Of all the factors affecting shelf life, most interest has focused on the possible difference in iced shelf life between fish caught in warm, tropical waters and fish caught in cold, temperate waters. In the mid- and late sixties it was reported that some tropical fish kept 20-30 days when stored in ice (Disney *et al.,* 1969).

This is far longer than for most temperate species and several studies have been conducted assessing the shelf life of tropical species. Comparison of the data is, as pointed out by Lima dos Santos (1981), difficult as no clear definition has been given on a "tropical" fish species and as experiments have been carried out using different sensory and bacteriological analyses.

Several authors have concluded that fish taken from warm waters keep better than fish from temperate waters (Curran and Disney, *1979;* Shewan, 1977) whereas Lima dos Santos (1981) concluded that also some temperate water fish species keep extremely well and that the longer shelf lives in general are found in fresh water fish species compared to marine species.

However, he also noted that shelf life of more than 3 weeks, which is often observed for fish caught in tropical waters, never occurs when fish from temperate waters are stored in ice.

The iced shelf life of marine fish from temperate waters varies from 2 to 21 days which does not differ significantly from the shelf life of temperate freshwater fish ranging from *9* to 20 days.

Contrary to this, fish caught in tropical marine waters keep for 12-35 days when stored in ice and tropical freshwater fish from *6* to 40 days. Although very wide variations occur, tropical fish species often have prolonged shelf lives when stored in ice as shown in Table. When comparisons are made, data on fatty fish like herring and mackerel should probably be omitted as spoilage is mainly due to oxidation.

Several hypotheses have been launched trying to explain the often prolonged iced spoilage of tropical fish. Some authors have noted an absence in development of TMA and TVN during storage and suggested that the spoilage of tropical fish is not caused by bacteria (Nair *et al.,* 1971).

The lack of development of TMA and TVN may be explained by a spoilage dominated by *Pseudomonas* spp.; however, qualitative bacteriological analyses must be carried out to confirm or reject this suggestion. Low bacterial counts have been claimed in some studies, but often inappropriate media have been used for the examination and too high incubation temperatures (30°C) have not allowed the psychrotrophic spoilage bacteria to grow on the agar plates.

Reviewing the existing literature on storage trials of tropical fish species leads to the conclusion that the overall sensory, chemical and bacteriological changes occurring during spoilage of tropical fish species are similar to those described for temperate species.

Psychrotrophic bacteria belonging to *Pseudomonas spp.* and *Shewanella putrefaciens* dominate the spoilage flora of iced stored fish. Differences exist, in the spoilage profile depending on the dominating bacterial species.

Shewanella spoilage is characterized by TMA and sulphides (H_2S) whereas the *Pseudomonas* spoilage is characterized by absence of these compounds and occurrence of sweet, rotten sulphydryl odours. As this is not typical of temperate, marine fish species which have been widely studied, this may explain the hypothesis that bacteria are not involved in the spoilage process of tropical fish.

Despite the different odour profiles, the level at which the offensive off-odours are detected sensorially is more or less the same. In model systems (sterile fish juice) 10^8-10^9 cfu/ml of both types of bacteria is the level at which spoilage is evident.

The relatively high post*mortem* pH is one of the reasons for the relatively short shelf life of fresh fish as compared to, for instance, chill stored beef.

It has been suggested that tropical fish species, such as the halibut from temperate waters, reach a very low pH, and that this explains the longer shelf life. However, pH values of 6-7 have been found in the studies of tropical fish species where pH has been measured (Gram, 1989). As the differences in skin properties are believed to contribute to the longer shelf life of flatfish, it has been suggested that this factor explained the extended shelf lives. It is indeed true that fish from warm waters often have very thick skin, but no systematic investigation has been carried out on the skin properties.

As spoilage of fish is caused by bacterial action, most hypotheses dealing with the long iced shelf life of tropical fish species have centred around differences in bacterial flora. Shewan (1977) attributed the long iced shelf lives to the lower number of psychrotrophs on tropical fish.

However, in 1977 only a very limited number of studies of the bacterial flora on tropical fish were published. During the last 10- 15 years several investigations have concluded that Gram-negative rod-shaped bacteria (e.g., *Pseudomonas, Moraxella* and *Acinetobacter)* dominate on many fish caught in tropical waters (Gram, 1989; Surendram *et al.*, 1989; Acuff *et al.*, 1984).

Similarly, Sieburth (1967) concluded that the composition of the bacterial flora in Narragansett Bay did not change during a 2-year survey even though the water temperature fluctuated with 23°C on a year-round basis.

Gram (1989) showed that 40-90% of the bacteria found on Nile perch were able to grow at 7°C. The number of psychrotrophic bacteria is within one log unit of the total count, and the level of psychrotrophic organisms is not *per se low* enough to account for the extended iced storage lives of tropical fish; Jorgensen *et al.* (1989) showed that a two log difference in number of spoilage bacteria only resulted in a difference of 3 days in the shelf life of iced cod.

The bacterial flora on temperate water fish species resume growth immediately after the fish have been caught and rarely is a lag phase seen. Contrary to this, Gram (1989) concluded that a bacterial lag phase of 1-2 weeks is seen when tropical fish are stored in ice. Also, the subsequent growth of psychrotrophic bacteria is often slower on iced tropical than on iced temperate water fish. This is in agreement with Liston (1980) who attributed differences in shelf life to differences in bacterial growth rates.

Although a large part of the bacteria on tropical fish are capable of growth at chill temperatures, they will (as this has never been necessary) require a period of adaptation (i.e., the lag phase and slow growth phase). Gram (1989) illustrated this by investigating the growth rate at 0°C of fish spoilage bacteria that had either been pre-cultured at 20°C or at 5°C. For some strains, the same bacterial strain would grow more quickly at 0°C if pre-cultured at 5°C than if pre-cultured at 20°C. Preculturing was done with several sub-culture steps at each temperature.

Similarly, Sieburth (1967) showed that although the taxonomic composition of the bacterial flora in Narragansett Bay did not change with fluctuating temperature, the growth profile of the bacteria fluctuated following the water temperature. However, the adaptation hypothesis does not explain why some tropical fish spoil at rates comparable to temperate water fish.

It can be concluded that many factors affect shelf life of fish and that differences in the physiology of the bacterial flora are likely to be of major importance.

Off Flavours Related to Fishing Ground

Occasionally fish with off-flavours are caught, and in certain localities this is a fairly common phenomenon. Several of these off-flavours can be attributed to their feeding on different compounds or organisms. The planktonic mollusc, *Spiratella helicina,* gives rise to an off-flavour described as "mineral oil" or "petrol".

It is caused by dimthyl-B-propiothetin which is converted to dimethylsulphide in the fish (Connell, 1975). The larvae of *Mytilus* spp. cause a bitter taste in herring. A very well known off-flavour is the muddy-earthy taint in many freshwater fish.

The flavour is mainly caused by two compounds: geosmin (1a, 10ß-dimethyl-9a-decalol) and 2-methylisoborneol, which also are part of the chemical profile of wine with cork flavour. Geosmin, the odour of which is detectable in concentrations of 0.01-0.1 μg/l, is produced by several bacterial taxa, notably the actinomycetes *Streptomyces* and *Actinomyces.*

An iodine-like flavour is found in some fish and shrimp species in the marine environment. This is caused by volatile bromophenolic compounds; and it has been suggested that the compounds are formed by marine algae, sponges and Bryozoa and become distributed through the food chain (Anthoni *et al.*, 1990).

Oil taint may be found in the fish flesh in areas of the world where off-shore exploitation of oil is intensive or in areas where large oil spills occur. The fraction of the crude oil that is soluble in water is responsible for the off-flavours. This is caused by the accumulation of various hydrocarbon compounds, where particularly the aromatic compounds are strong flavourants (Martinsen *et al.*, 1992).

Bibliography

Adelberg, Edward A.: Papers on Bacterial Genetics, Boston, Brown and Company, 1960.

Bailey, Ronald: Liberation Biology: The Scientific and Moral Case for the Biotech Revolution, Amherst, Prometheus Books, 2005.

Barnes, N.: Biology, New York, Worth Publishers, 1989.

Bauer MW: Biotechnology-the Making of a Global Controversy, Cambridge, Cambridge University Press, 2002.

Brandwein, P.F.: Sourcebook for the Biological Sciences, San Diego, Harcourt Brace JOvanovich, 1986.

Broach, J.R.: The Molecular Biology of the Yeast, Cold Spring Harbor, Cold Spring Harbor Laboratory, 1981.

Chrispeels, Maarten : Plants, Genes and Crop Biotechnology, Sudbury MA, Jones and Barlett Publishers, 2003.

Conko, Gregory: The Frankenfood Myth: How Protest and Politics Threaten the Biotech Revolution, New York, Praeger, 2004.

Curtis, H.: Biology, New York, Worth Publishers, 1989.

DeGregori, Thomas R.: Bountiful Harvest: Technology, Food Safety and the Environment, Washington DC, Cato Institute, 2003.

Dodds, John H.: Plant Genetic Engineering, New York, Cambridge University Press, 1985.

Drexler, K. Eric: Engines of Creation: The Coming Era of Nanotechnology, New York, Anchor Books, 1986.

Fransman M, Junne G, Roobeek A: The Biotechnology Revolution?, Oxford, Blackwell, 1995.

Friedberg, E.C.: DNA Repair, New York, WH Freeman and Company, 1985.

Fumento, Michael: Bioevolution: How Biotechnology is Changing Our World, San Francisco, Encounter Books, 2003.

Gaskell G: Biotechnology-the Making of a Global Controversy, Cambridge, Cambridge University Press, 2002.

Goodman, D.C.: From Farming to Biotechnology: A Theory of Agro-industrial Development, Oxford, Blackwell, 1987.

Horn, Toby M.: Working with DNA and Bacteria in Precollege Science Classrooms, Reston, National Association of Biology Teachers, 1993.

Kurzweil, Ray: The Age of Spiritual Machines, New York, Penguin Books, 1999.

Miller, Henry I. : The Frankenfood Myth: How Protest and Politics Threaten the Biotech Revolution, New York, Praeger, 2004.

Moravec, Hans: Mind Children: The Future of Robot and Human Intelligence, Cambridge, Harvard University Press, 1988.

Murray, David: Seeds of Concern: The Genetic Manipulation of Plants, Sydney, University of New South Wales, 2003.

Old, R.W. : Principles of Gene Manipulation, London, Blackwell Scientific Publications, 1989.

Postman, Neil: Technopoly: The Surrender of Culture to Technology, New York, Vintage Books, 1992.

Retzer, W.J.: Biotechnology Workbook, Englewood Cliffs, Prentice Hall, 1991.

Rifkin, Jeremy: The Biotech Century, New York, Penguin Putnam, 1998

Shetty, Kalidas: Food Biotechnology, New York, Dekker/CRC Press, 2005.

Smith, John E.: Biotechnology, Cambridge, Cambridge University Press, 2004.

Switzer, R.L.: Experimental Biochemistry, New York, W.H. Freeman and Company, 1977.

Towle, Albert: Modern Biology, Austin, Holt, Rinehart and Winston, 1988.

Walden, Richard: Genetic Transformation in Plants, England, Open University Press, 1988.

Index

□□□